The
Human Being
and the
Animal World

The Human Being and the Animal World

Charles Kovacs

Floris Books

Publisher's Note

The Human Being and the Animal World is based on extensive lesson notes written by Charles Kovacs in the 1960s and reflects the author's interpretation of the subject. We recommend reading this book with an understanding of the context in which it was written.

First published in volume form in 2008
Fifth printing 2023
© 2008 Estate of Charles Kovacs
The author has asserted his right under the
Copyright, Design and Patents Act 1988
to be identified as the Author of this Work
All right reserved. No part of this publication may
be reproduced without the prior permission of
Floris Books, Edinburgh
www.florisbooks.co.uk

 Also available as an eBook

British Library CIP Data available
ISBN 978-086315-640-3
Printed and bound in Great Britain
by Bell & Bain, Ltd

Printed on sustainably sourced FSC® certified paper. Uses plant-based inks which reduce chemical emissions.

MIX
Paper | Supporting responsible forestry
FSC® C007785

Contents

Foreword

The contents of this book are taken from Charles Kovacs' notes for the "Human Being and the Animal World" topic, which is usually given in Waldorf schools as part of the curriculum for Class Four, for nine- to ten-year-old children. Since he was teaching in Edinburgh, the choice of animals has a mainly Scottish theme and teachers in other parts of the world would probably choose to emphasize their own local fauna. Part Two was given in the next school year and unfortunately the end of the manuscript has been lost. For the sake of completeness — and as a former pupil of Mr Kovacs — I have taken the liberty of adding the chapters on the lion and on Buddy the guide dog. Of the art work which originally accompanied the main lesson period, three songs have been included.

Charles Kovacs follows the suggestion made by Rudolf Steiner in speaking of the animals as head-animals or trunk-animals.★ For a thorough discussion of the science behind this, one could hardly do better than refer to the book, *Man and Mammals,* by Wolfgang Schad and the article "Seeing the Animal Whole" by Craig Holdrege. The present volume is in no way intended as a textbook which ought to be followed — rather the hope is that it might act as a stimulus to the creativity of individual teachers working with their own particular classes.

Howard Copland

★ Rudolf Steiner, *Practical Advice to Teachers,* Lecture 7.

Introduction

From a talk given by Charles Kovacs to the parents of his class.

Quite a big part of the term was taken up by the "Human Being and the Animal World" topic, and I would like to explain the importance we attach to it. Of course on the plainest level one could say: we tell the children about animals at this age simply because they want to hear of them. It is quite an elementary fact that children show an enormous interest in animals. This care for animals often takes quite astonishing forms — I once heard a story which is quite typical of the child's attitude to animals. A father took his little boy to a picture gallery where there was one of these large historical paintings showing Christian martyrs thrown to the lions in a Roman circus. The father explained to the boy what the picture was about and spoke at length of the heroism of the martyrs. But he was suddenly interrupted by his offspring who pointed excitedly at one corner of the picture: "Look daddy, look, there is a poor lion who hasn't got a martyr!"

But now we must make a distinction between the interest in animals which children have before the age of nine or ten and the quite different interest which grows in them when it comes to Class Four. Younger children see and feel the animal rather as something that belongs to the world of the fairy story; they will not regard it as extraordinary if they are told of animals talking and acting as if they were human beings. This is no longer so for children between nine and ten years of age; now they awaken to the fact that animals are essentially different from people. In short we now have the beginning, the first realization of the world of nature as distinct from the world of human beings.

Now what has happened in the mind of children to bring about this change?

I would like to explain this change as clearly as possible, first of all because it can be a key to the whole mentality of children at this age, but also because an understanding of this change also illuminates the working of adult minds. Now we adults make necessarily and naturally a very clear distinction between two mental functions: rational, logical thought, and imagination. We have to keep these two functions well apart and we would be considered mentally abnormal if we could not do so. But, like two branches, these two functions grow from a common stem, and in the stem they are not divided, but are one. And this division which we take for granted today is, historically speaking, no older than Greek philosophy and Greek science, that is about 500 BC. Before this time even adults had a mind or a mentality in which rational thought and imagination were not two different things, but one. It is no wonder then that these earlier ages had no philosophy and no science as we know it; they had mythologies; they regarded nature not as a modern scientist does, but as filled with beings, gods and monsters.

To some extent children below the age of nine or ten still live in this stage where the rational and the irrational, imaginative, element of the human mind are not divided. For young children the statement that "a dragon threatened the princess" and the statement that "two times two equals four" are not of different orders, they are taken in as realities of the same kind. And if we induce young children — as we can — to make a sharp distinction between the two, then we do irreparable harm to them. We force upon children a speeding up of mental development which will show itself as emotional disturbances at puberty. However, between nine and ten the division begins naturally; now the faculty of rational thought begins to branch away from imagination. And this is the reason why children now want some food for their intellectual capacity and why they want to hear about animals not in the form of fables and fairy stories, but in the form of more objective descriptions. In fact children want to be introduced into the world of science.

This is a very important moment when we lead children to the portals of science. And now something must certainly be avoided for children of this age: that we throw at them the vast conglomeration of theories and hypotheses which are so often mistaken for "established" scientific truth. And this is particularly the case in relation to the animal world. Ever since the late nineteenth century, many people have mistakenly regarded the animal kingdom as a world ruled only by a ferocious struggle for the survival of the fittest. Many of the books written for children are but watered down forms of this view, in spite of the fact that as a serious scientific theory this over-simplified form of Darwinism is already a thing of the past. But if, directly or indirectly, this picture of the animal world were impressed upon children, we would — whether we want it or not — implant into them the seeds of brutal and ruthless egotism; we would educate the amoral and not the moral element in children.

Thus we try to teach children to look at the animal always in comparison to human beings. We emphasize not at all that we have a better or more highly developed brain than the animals, but that the thing which lifts us above the beasts is our upright walk, which frees our hands, and that by the things we do with our hands we serve not only our own needs but also the needs of our fellow human beings. In this way we can bring children to an objective, rational understanding of nature, but at the same time strengthen the social and moral element in them. So you see that the "Human Being and the Animal World" topic is not merely introduced to meet the natural interest of children, and not merely for the sake of instruction, but as an important help in their moral education.

Earlier we mentioned the two functions of the mind: rational thought and imagination. I need hardly stress that we do not regard imagination or fantasy as an inferior function and that we want to provide food for it as much as for the other. In Class Four we do this by telling the children the stories of Norse Mythology, the great sagas of Thor and Odin. These stories may appear grim to you, but the children simply love them. Why? Because they feel something in them which will bear fruit only much, much later. In these sagas the everlasting struggle between good and evil in the human soul is, as it were, projected into the cosmos and that which is good is shown not as a dry moral precept but as beautiful beings, the Asas, and evil is shown as monstrous ugliness. Here again we try to work from the moral element; the children learn to love good as something beautiful and to hate evil as something ugly. And these ancient pictures have a much stronger moral force than any preaching of morals could ever have.

Having spoken of the awakening of rational thought at this age, I must mention something that arises in children as a consequence. We adults use our rational thinking not exclusively for the purpose of knowledge; we use it often enough, rightly or wrongly, to criticize others. The faculty of criticism is simply the by-product of being able to think rationally. And so we have to be aware that these children will no longer accept the authority of their parents or of their teachers unquestioningly. They now begin to be aware of our personal weaknesses and shortcomings. And there is much more required, both from parents and teachers, if we wish to remain what we have to be: our children's guides into life. We have to be aware of the fact that these children assume — quite naturally and instinctively — the role of little judges, and they will not let us off lightly, for we shall lose our authority if we do not live up to the standards they expect from us.

Another side of this new clarity which rational thought gives to children is an extremely keen sense of fairness. Children are, one could say, oversensitive to anything they consider unfair. One day I told the children in my class a story which was really intended to make them work out some fractions. It is the story of a ring lost by a king: he offers a great reward to anyone who brings the ring back to him. A soldier, a mere private, finds the ring and decides to bring it to the palace personally. Naturally he has to pass several officials before he can see his majesty and each of the officials extracts from the poor soldier a promise to give him a share of the reward, so that by the time he comes before the king his whole reward has already been forfeited to the greedy officials. You should have seen the faces of the children at this part of the story, how they registered profound resentment at the utter unfairness of the proceedings, and how their faces lit up when they heard what the soldier answered when the king asked what reward he wanted. He said he wanted 120 strokes with a whip, and this reward was duly distributed among the officials. I am afraid that the children were so pleased with the justice of this ending that they had, by then, quite forgotten to work out the fractions.

Part I

1

The Head, Trunk and Limbs

Now that we are getting cold days we all must see to it that we are well dressed, that we are clothed and covered against the cold. We put on hats and caps, we wear overcoats, we wear long trousers or tights, we put strong shoes on our feet and we put our hands in warm gloves. But think now how different all these things are which we need to be well covered. A cap is quite different from a glove and it would not be much good if you put a glove on your head and a cap on your hands. You see, we need so many different things because our body — the human body — is itself made of so many different parts.

A cap or a hat must be round because our head is round. Our head as a whole is really like a globe or a ball, except the part where it sits on the neck, it is a bit flattened there, but otherwise the head is like a globe, or a ball; it is called a sphere. And if I wanted to make a head in wax or plasticine, I would start, before I make the face, by making a little ball. Or think of the Hallowe'en lanterns made from turnips or pumpkins which are round, and so by cutting some holes in them they can look like heads. You could hardly make a round turnip into a leg or an arm! So in a way the head is easy to model; it is round, as round as the sun, as round as the full moon. When you look up to the full moon it is really quite easy to imagine that there is a face up there looking down, because the full moon is round; it is a perfect circle.

That's the top part of our body, the head. But now we go a bit lower, the chest and the belly part, which is called the trunk. The part of the body without head, arms or legs is called the trunk. A vest or a waistcoat covers just the trunk part of us.

While the head is like the full moon, the trunk is not like the full moon, but rather like a half moon. It would not really be right to make the trunk as fully rounded as the head. When we make snowmen, we *do* make the trunk out of a big round ball of snow like the head, but it is not really like a true human body; we must hollow it out, a bit at least, to make it more like the real trunk. Even a banana — if it is not curved too much — is more like a trunk than a fully rounded ball would be. If we put a round ball of wax or plasticine on a banana, it would look like a head on a trunk. So the head is like the full moon, but the trunk is more like the half moon.

And now we come to the third part of the body, the limbs, the legs and the arms, the parts that are covered by trousers and tights and sleeves. And you only have to look at trousers and sleeves to know that the limbs are straight; just as the rays that come from the sun or moon are straight. If I put some sticks into the plasticine trunk you would recognize them as arms and legs.

The limbs, arms and legs, are straight. We can bend them of course but when we stand straight, the arms and legs are straight. The trunk cannot be quite straight (it bends a little in or out, because it is nearer the half moon shape) and the head will always be round. And now you can see why we have such different kinds of clothing: because they have to fit such different parts of the body.

These are the main parts of the body: the round head, the half-moon shaped trunk and the straight limbs. The greatest difference is between the head and the limbs.

Just think of all the things the limbs do. First of all our legs — they are really the faithful servants of the body — carry the whole weight of the trunk, the arms and the head. And yet, carrying all this weight, they can not only stand, but can walk, run, leap, jump and dance. If you make a wax or plasticine figure it is very difficult to make them stand on two slender legs; as our real legs are. But our real legs carry all that weight quite easily, and we not only stand, but walk or run with them. They are very good and very strong servants of the body.

Our arms and hands *can* carry weight, but they don't *have* to as the legs do. Our arms and hands are free to do so many things that we could not name them all: draw, paint, write, play instruments, play with balls, with toys, with dolls, lift things and put them down, steer cars, ships, planes, take a little child by the hand, shake hands when we greet each other, help somebody up who has fallen. Think of all the things a doctor does, a nurse does, a gardener does; there is no end to all that hands can make and do. So, of our limbs, the legs are quite wonderful servants, but they can only do a few things. Our arms and hands can do endless things. Our hands are free to do one thing or another, while our feet can do only one thing, carry the body, but they do it very well.

And compared with the hands and even the legs, the head is a very lazy fellow, sitting comfortably on the neck and doing very little work. He sees and hears and smells, he watches what is going on but he leaves all the real work to arms and legs. Of course we need this round fellow, the head, but we also need to know that he is the one part of our body that does the least real work; the hardest work is done by the limbs. And the trunk is just in between; the trunk does some work, for in the trunk there is the heart that beats, the lungs that breathe, the stomach that digests our food. So the limbs are the most active part, the trunk a little less active and the head is the least active of them all. We have now learned something of our body, of its three parts: head, trunk, limbs. And we now know that most of the work is done by the limbs and that the hands are the most wonderful part because they can do more than all the others.

★

We looked at the three different parts of the human body — the head, the trunk, the limbs — and we saw a great difference between the two kinds of limbs, the legs and the arms. The legs are servants and the hands are free to do so many things that we just could not count them all.

The legs are good, strong and faithful servants that we can trust much more than a car. But these good servants, the legs, serve only our own body. But if your mother uses her hands to make a meal for you, what the hands do is not only for herself, it is for the whole family, it is for others. If I draw a picture for the class, my hands do something for you, not for myself. Think of the work done by builders, by farmers, by fishermen, by miners. They all use their hands for things which other people need. And so we can say: the legs and feet are our servants, which serve our own body, but the hands serve the world; they serve others. Of course we often use our hands for ourselves, but we can also use them for others. When somebody falls and we help them up, our hands serve others. And now we can begin to understand how wonderfully the human body is made: the head that sits still and the limbs that work; but the legs work for ourselves only, the hands can work for others, for the whole world. And between the still head and the working limbs there is the trunk.

The head is not just a round ball; in the head are our eyes, our ears, our nose and our mouth. All the things we see, all the sounds we hear, all the smells and every taste: sweet, sour, bitter, salt, it all comes to me through my head. We say something is big or small, beautiful or ugly, but we know this only because we have eyes in our head. We hear what people say, we hear music or a poem, the noise of traffic or the sounds of animals; but only because we have ears in our head. And, as a little child, we could not have learned to speak if we had no ears that could hear. The smell of a good dinner, the smell of roses, the smell of a cow, all these smells which tell us so much about the world, we know them only because we have a nose in our head. And we would not enjoy our food much if we could not taste it in our mouth. Just think what a poor life it would be if everything we ate had the same taste.

Most of what we know about the world comes to us through the head. The head does not work hard (as the legs or the arms and hands do) but it has, one could say, doors through which the whole world comes to us. And when we compare the hands with the head, what can we say? With the hands we give something to the world, all the things we make with our hands, they are given to the world. But through the eyes, ears, nose and mouth in the head, the world gives something to us; it gives us colours, shapes, sounds, smells, tastes; and all our knowledge. So we can say, the hands *give* to the world, the head *receives* from the world, it is given all the sights, sounds, smells, tastes.

While the head has these openings to the world — eyes, ears, nose, mouth — what comes in through these openings does not stay in the head. If we have something nice in our mouth and we taste it, it does not stay in our mouth, it goes into the trunk, into our stomach to be digested. We breathe in a nice smell of flowers, but the air we breathe, the air with the nice smell, does not stay in our nose, it goes into our lungs which are in the trunk. And even with sound it is similar; if we hear a horrible noise, we can feel how it goes right down into our body, into the trunk, while a nice tune makes us feel well in our heart. And so it is with the beautiful or ugly things that we see; we can feel them right down to our heart that is in the trunk. The head receives all the sights and sounds and smells and tastes of the world, but they go farther, they go into the trunk.

So the human body is something quite complicated. If we start below we must say that the legs serve ourselves, the hands serve the world. Then we compare hands and head: the hands give to the world, the head receives things from the world, like sights, sounds, smells, tastes. But these things don't stay in the head. They go down to the trunk. And what happens to all that comes into the trunk? Think of the food. It is digested in the stomach, and when it is digested it makes our limbs, legs and arms strong to do all the work they have to do. All that is beautiful, nice, good, it makes us strong. That's how it is, the world gives us our food, and the food makes us strong so that we can give something to the world. Through the head we receive, through the hands we give.

The Cuttlefish

Having heard about the human body we will see that we cannot understand the different kinds of animals in the world unless we first know of the three parts of the human body — the head, the trunk and the limbs.

All the many, many kinds of animals that fly and run and crawl and swim can be divided into head-animals, trunk-animals and, perhaps, limb-animals. When I say head-animals, I don't mean that they have only a head. Of course there is no such animal! They all must have trunks and limbs to live. But head-animals are animals which behave, as a whole, like our

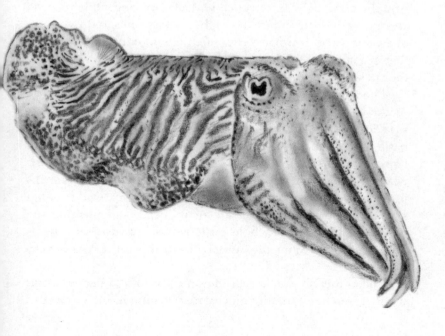

head behaves by itself. The whole animal behaves as if it were just a head. How does our head behave? The head does not work like legs or hands, the head watches what is going on around it; as we said, it takes from the world. The head, one could say, "sucks in." We suck in air through the nose, food through the mouth, but we also suck in sound through our ears, the sound has to go right into the inner ear. And even if we look at something, something beautiful, for instance, it is as if our eyes want to get hold of it, want to draw it into ourselves. That is how the head behaves.

There is a curious creature that lives in the sea, called a cuttlefish. It is not a fish at all, nor does it even look like a fish, but it does live in the depths of the sea. It is not a beautiful creature but a very strange one. This cuttlefish has two very big eyes that are large in proportion to the body. The body is about 40 cm (15 in) long but the eyes are the size of cherries. Under the eyes, which bulge outward on either side, the cuttlefish has a beak like a parrot, but you can hardly see it because around the beak there are tentacles. Tentacles are kind of arms without bones that twist and turn like snakes. And the cuttlefish has ten such tentacles or long rubbery arms around its mouth. There are eight shorter ones and two long ones, and each arm has a row of things looking like buttons, that are suckers. If the arms or tentacles touched your skin, the suckers would stick.

The front part of the cuttlefish has eyes, beak and tentacles. At the back there is a kind of trunk covered with a loose skin that does not fit the body tightly. The skin (called the mantle) has flaps on either side of the body and the cuttlefish uses these flaps like fins when it swims slowly. But when the cuttlefish wants to move faster, it does something quite different. It sucks in water through its gills and pushes the water out through a special spout on the underside of its head, and pushing the water away pushes the whole cuttlefish backwards. So by pushing out a jet of water the cuttlefish can move backwards very quickly.

The cuttlefish watches all the time with these big eyes, and while it watches, it hardly moves, just floating in the water. But when it sees something it wants — a little shrimp, for instance — it sucks in water, pushes it out and is near the shrimp in a moment. Then the long arms shoot out to get hold of the shrimp. They bring the shrimp to the shorter arms, and now all the arms with their suckers hold the shrimp fast. Then the cuttlefish has its breakfast.

Now *our* eyes haven't got tentacles with suckers to grasp hold of things we like, but if you watch some greedy people looking at something they want, it is as if there were invisible tentacles getting hold of this thing, holding on to it with suckers: they can't let it go from their sight.

★

If you think of the cuttlefish floating or moving itself by a jet of water from the spout, it is as if this strange creature were a kind of head that swims about by itself. As the head is watchful of all that goes on around it, so the cuttlefish watches all that goes on; it is very alert and notices everything that goes on in the water around it.

What is more, the cuttlefish has a very special way of showing that it notices everything around it, for it can change the colour of its skin or mantle in a quite marvellous way. If it is floating over some yellow sand at the sea bottom its skin becomes orange-yellow; if it floats over or between seaweed with a brownish green colour, the cuttlefish turns brownish green. In a dark cave it turns black, and in clear water it becomes so light green that you can hardly see it. What happens is this. The cuttlefish sees the colour of the seaweeds and the colour goes not only to its eyes but right into it and shows itself in the skin's colour like a kind of photograph or film of whatever is around the cuttlefish.

Of course, it is also useful for the cuttlefish that it can change its colours so that it looks like the seaweeds around it. Its enemies, like dolphins and porpoises, who like to eat cuttlefish can't see it so easily if it is coloured like the things around it. We human beings learned to use this trick only a short time ago; soldiers fighting in a forest are given dark green uniforms, for instance. This trick is called camouflage, but the cuttlefish could do it long before we thought of it.

The cuttlefish also changes its colours for other reasons. If it is floating over sand, its colour is orange-yellow. But suddenly it sees a little shrimp, and it feels just like we feel when we are very hungry and a nice meal is put right in front of us. We don't show it, of course, even though our mouth may be watering, but the cuttlefish does show how much it wants to tuck in and have its dinner. The skin on its back becomes coppery red, the tentacles turn green, the eyes gleam pink and blue, even the belly changes colour, yellow, pink, purple, red. And then the long tentacles reach out to catch the shrimp, pull it in, and the cuttlefish gobbles its prey. Then its colours go back to normal, that is, whatever colour is around it.

If we human beings see something we would like to eat, it does not show. But if the cuttlefish sees something it wants, it shows it right away in the colours all over its body. There are also things we don't like, or things we fear, and again, we don't change our colour much when we are afraid. Perhaps we get a little pale, but that's all. The cuttlefish does something quite different. When it feels there is something approaching that could be a danger, it does not change the colour of its skin, it shoots out a liquid as black as ink, and in an instant there is a black cloud in the water. Leaving the black cloud behind, the cuttlefish swims as fast as it can to a safe place, as it takes quite some time until the black stuff is dispersed. It is fear that makes the cuttlefish throw out a black cloud into the water.

So we can see how keenly the cuttlefish is aware of all around it: it takes on the colour of its surroundings, it changes its colours when it sees something it wants to eat, it throws out a cloud of ink when it feels threatened. The cuttlefish is alert with its whole body, while we are mainly alert with our head. That's why the cuttlefish is like a head. It is a head-animal.

In olden times fisherman caught cuttlefish for that black ink inside. It was really used as ink, though it actually is a special kind of brown, known as "sepia." Most cuttlefish caught today are caught for something different. The cuttlefish has only one bone in its body, a flat bone called the cuttle-bone. People feed this to cage birds such as parrots, to strenthen their beaks and bones.

3

The Seal

When we think of all we have heard about the cuttlefish, we can easily see that the cuttlefish is mainly a head: it has no limbs at all, not even fins like fish, and it must use his skin to swim, or spout jets of water. The tentacles around the mouth, which it uses as we use hands to get hold of something, are really like tongues, so there is just the head: eyes, tongue, mouth, and barely a trunk stuck onto it. It is so aware of everything around it that it changes its colour. The cuttlefish is a head-creature. It is more head than trunk or limbs.

Now we shall look at a trunk-animal, an animal which is more trunk than head or limbs. In such an animal the trunk would be the most important part, perhaps even the largest part. The head would be just a continuation of the trunk, as if head and trunk were one piece, and the limbs would not be of any great size either. Such an animal is the seal.

It is easy to draw a rough outline of a seal and this outline shows you straightaway that the seal is mainly a trunk, while the head and the limbs are small in comparison. If a creature has such a big trunk and small limbs, it would be a bit like a beached whale when it is on land. The seal is really not at home on land, it is at home in the water, in the sea. Seals do go on land sometimes, but they can't move well or fast, they have to crawl a bit like caterpillars using the trunk itself to wriggle forward. Their limbs, which are flippers, are used on land. But of course, in the water the flippers are wonderful, the hind flippers drive the seal through the water with great force and great speed, and the front flippers give the animal balance.

If we look at the shape of the whole body, trunk and head, we can see that it is just the right shape for moving in water. When we human beings build a rowing boat we put the narrow end in front and the blunt end at the back. The narrow or sharp end can cut through the water much better, it would be much harder work if we tried to row in the direction of the blunt end, and the body of the seal has just the right shape, the narrow end with the head in front. The seal's body is streamlined, which means it has a shape that makes the water stream smoothly past it.

But in other ways too the seal is wonderfully built for life in the water. It is not nice for us human beings to get water in our eyes or in our nose. The seal does not have to worry about this: the ears, just openings, have a strong skin that closes them when the seal dives, and the nose too closes when the seal dives. The seal does need air, it breathes air just like us, but when it dives it breathes out. This may seem strange, but if the seal had a big lungful of air it would find it hard to get beneath the surface of the water. It then swims underwater for as long as half an hour before it has to come up for breath. No human diver could do that. Another strange thing is that when we dive and hold our breath our heart beats faster, but the seal's heart beats slower. The seal stays underwater without any new air in its lungs and sometimes even sleeps for a little while underwater.

There is another thing that makes life in the water easy for the seal. All through the winter when the water is icy cold the seal is in the sea. We would freeze to death in a few minutes because our blood is warm and cannot stand great cold. The seal also has warm blood but under the skin it has a thick layer of fat, called blubber, and this thick layer of fat keeps the cold out and keeps the warmth of the blood inside.

So the flippers, the shape of the body, the skins that close nose and ears, the lungs that let the seal stay under water for a long time, the thick layer of fat, they all make the seal a water-animal.

★

There are different kinds of seals. On our Scottish shores there are two kinds, grey seals and brown seals, but the grey seals are often dark brown and the so-called brown seals are really dark grey. And when they are wet and their coats are shiny you can hardly tell what colour the fur is. Yet there is one thing by which you can always tell one kind from the other: the grey seals have more pointed noses and the brown seals have snub noses.

We shall look at the life of a seal from the time it is born. We shall first look at the life of the grey seal. Human beings are born at any time of the year, people can have their birthdays in any month of the year, but it is not so with wild animals, and grey seals are born in October or November. Little seals are called seal pups and they are born on land. The seal mothers come often in their hundreds to an island or to a beach to give birth to their young. In these places at that time there are really crowds of seal mothers and seal fathers (the female seals are called cows and the male seals bulls, just as cattle are). There they wait for the babies or pups to be born. Each seal mother has only one pup, not like dogs or cats who have many young ones.

When a little seal pup is born, though it is going to be a grey seal, its fur is white for the first three to four weeks, then it gets darker. And this newborn seal pup is at first quite helpless, it lies on the sand or rock and can just lift its head, but cannot yet move itself. This is the time when the little one is in great danger from gales and storms, for a big wave thrown up by a gale can sweep the pup into the sea and then it is lost. But the mothers protect their little ones by lying on the seaward side of the pups.

The pups feed happily on their mother's milk. Now the milk of the seal cow is much richer in fat, in cream, than cow's milk, and that is why the seal pup grows at an amazing rate, gaining 1kg (3 lb) every day. The mother does not stay with the pup all the time; she has not eaten for a whole week and so must go and catch fish in the sea. When she is away the pup cries desperately, and then you can see something that is very strange, for seals are the only animals that cry tears like we do. The little puppy yells and whines and tears run down its face. But after a time, of course, the mother comes back. On that beach there are hundreds of little pups who all look the same to us, but the seal mother knows which one is her puppy, and never makes a mistake.

As the seal pup grows so fast, it can soon move about on land and it begins to play with the other puppies. A favourite game is that it hits another puppy with a flipper and then bounces away as fast as it can, and the other pup tries to catch it. Or they even have wrestling matches, or they splash in pools between the rocks. But the grey seal pups don't go into the sea for a swim, in fact they are scared of the sea, they can't swim yet, and when a big wave comes they bounce away as fast as they can. When they are about two weeks old, their mothers take them into the water and, at first, the puppies are not pleased at all. They cannot tread water, and would sink if their mother did not hold them up with her fore-flippers, or sometimes they ride piggyback on their mothers. It takes only three to four days for the seal pup to lose its fear of the water, and learn to swim, and to close its nose and ears, and to breathe out when it dives down. And now the young seals have a lovely time in the water, they slip down the steep side of a wave and ride to the top of the next wave to slide down again. They dive and swim over and under each other, they twist and turn and race each other, and it all shows that they have become really at home in the sea.

★

Grey seal pups are at first not at all at home in the water, and they cannot swim. There was once a sailor who found a baby seal washed ashore on the beach by a gale. He took the little thing and, out of kindness and because the sea was calm again, he threw it into the water. But the seal just sank like a stone and the sailor had to jump in and dive for it to rescue it from drowning. The seal was not yet old enough to swim. Then the sailor tried to feed it, he cut up some fish, but the baby seal would not touch fish and so he had to feed it on milk from a bottle. But even then the seal baby would only take the rubber teat if the man first put a lot of butter on it. But, in the end, this seal too learned to swim.

When the seal pups are three weeks old they have grown

from 15 kg (30 lb) at birth up to 50 kg (100 lb). They are sleek and fat from their mothers' rich milk, their fur has changed from white to speckled grey-brown. And now that they can swim the mothers and the fathers leave them, swimming out into the sea, while the young ones remain behind on the beach.

Just think how long human parents look after their children, for years and years. But the seal parents leave their young to fend for themselves when they are only three weeks old. This is a hard time for the young seals; their only food has been the mother's milk. For a week or even ten days the young seals eat nothing, they just lie on the beach or on a rock. But this fasting is good for them, for in this time the nourishment in their bodies, all the rich milk they have had, turns into firm strong muscles. And when, at last, they go into the sea they are strong and fast, and they soon get used to living on fish.

And once they have left the shore or island where they were born, the young grey seals — not the older ones, but the first year seals — are great wanderers, they swim for hundreds of miles. Seals from Scotland may go as far as Spain or Norway. For weeks and months they do not come ashore, but they live in the sea, feed in the sea, and sleep in the sea. And all this swimming and wandering they do in the coldest time of the year — December, January, February, right into spring. After about six months they swim back to the place where they were born, and no one knows how these young seals can find their way back over hundreds of miles. But they do come back, and when it gets warm and sunny they lie along the beaches and rocks where they came into the world.

All this is about the first year or so in the life of a grey seal. With the brown seal it is a bit different. Brown seals are sometimes born on land, but often in the water. Brown seal pups can swim from the day they are born. They can be fed by their mothers underwater and don't mind a bit. And from the first day, the mother plays all kinds of swimming games with the little one: she dives with it, holding it in her fore-flippers, she helps it through rough waters and big waves, and she also helps it to go ashore, for the pup's flippers are not strong enough to pull it up. But brown seal pups spend most of their puppy time in the water. With the brown seals too, the mother leaves them after three to four weeks and then they are on their own.

Miss Farr's pet seal

We have heard about the young grey seals and brown seals that live along the coast of Scotland. An interesting thing about these creatures of the wild sea is that if they are caught quite young they can become greatly attached to human beings. They can become as fond of you as a dog would be. I'm going to tell you about a brown seal (brown seals are really clever and also friendlier to people than grey seals).

One such person who kept a young brown seal was a Scottish lady, Miss Rowena Farr, who lived with her aunt on a croft in Sutherland. Miss Farr was very fond of animals; she already had two otters and two squirrels, and then she went to visit some friends who lived on the Hebridean Islands. And while Miss Farr was with her friends she met a fisherman who in his arms held a little furry bundle, a seal baby that had been washed ashore by a gale. Miss Farr immediately asked if she could have it, the man gave it to her, and so Miss Farr came into possession of the little seal puppy.

Her first problem was how to feed it and on what. People told her to feed the seal with warm milk mixed with oil, the oil was to make the milk as rich, that is as fatty, as a seal mother's milk. The seal puppy took to the milk bottle from the first day and enjoyed it. And a few days later Miss Farr returned to her and her aunt's croft in Sutherland carrying a bulky parcel weighing more than 15 kg (30 lb) — the seal. She had already decided on its name: Lora.

The upbringing of Lora was not easy. Lora had come to regard Miss Farr as her mother — after all she gave her milk — and so whenever Miss Farr could not give all her attention to Lora because she had other things to do, Lora started to make the whining noise of young seals and cried tears. In the first few days Miss Farr had put Lora on her lap while feeding her from the bottle. Now that was alright when the creature weighed only 15 kg (30 lb), but Lora grew in size and weight, and even when she was 45 kg (100 lb) she still wanted to go on Miss Farr's lap and barked and yelped when she was not allowed to.

And as soon as Lora was strong enough to move about she waddled after Miss Farr, following her wherever she went. Well, Miss Farr liked to go for long walks and the seal could not keep up with her fast walking pace, and so as soon as Miss Farr was some distance ahead Lora set up a heartrending barking and yelping to call her back. So Miss Farr decided she must find some way for Lora to amuse herself. There was a loch nearby and on it Miss Farr had a rowing boat. She put Lora in the boat, rowed out and threw Lora overboard. Lora, being a brown seal and not a grey one, was immediately at home in the water — diving, twisting and circling the boat with incredible speed.

From now on Lora took to the water of the loch every day, enjoying herself and also finding at least part of her food for herself by catching fish in the loch. But she still liked the extra bits she could get in the house: dog biscuits soaked in milk, raw carrots, even porridge — and as a special treat she was often given the oil from sardine tins. Lora also had company when she went for a swim in the loch, two otters, and she and the otters became playmates and swam and dived happily together. But Miss Farr had only to call her name and Lora came out of the water immediately, like any well-trained dog. Lora also learned very soon that when she came from the lake, all wet, she could not waddle straight into the living room. There was an old mackintosh on a long shelf at the entrance, and Lora used to pull it off the shelf and sit on it until she was dry. Then she barked to be let into the living room.

But though she was well-trained in this way, Lora could be very naughty too. One day she took the tablecloth between her teeth and pulled hard, bringing down plates, glasses and cutlery in a heap. On the other hand Lora liked to make herself useful. After a time she recognized the postman and when she saw him coming she would waddle to meet him, take the letters in her mouth and bring them into the house. But on one occasion she decided, while having the letters in her mouth, to go first for a swim in the loch. Well, Miss Farr never got this bundle of letters.

★

Miss Farr wrote about all her experiences with her animal pets in a book called *Seal Morning* and you can read much more about Lora than I can tell you here.

People who live on the seashore and get to know a lot about seals have always said that seals are attracted by music and singing. And this, as it turned out, was quite true with Lora. There was a piano in Miss Farr's croft and when Miss Farr sat down and began to play, her other animals took no notice. But Lora would come, lean against the piano legs, or against Miss Farr's own legs, and listen with intense concentration, and Lora's face then always had an expression of great joy. She would also sway her body in time with the music.

Playing the piano was alright, but one day Miss Farr wanted to play and to sing a song. As soon she started to sing there came a groaning noise from Lora. Miss Farr was not going to be interrupted, she went on singing, but Lora kept up her own singing too. And, of course, Lora's voice was so much stronger: she roared, mewed and wailed so loudly that Miss Farr could not hear her own voice. So Miss Farr stopped singing but continued to play the tune on the piano. By now Lora was enjoying hearing her own voice and wailed happily and loudly to the tune played on the piano. There was not much tune in Lora's roaring and wailing but after much practice with Miss Farr, one could just recognize something like ba-ba black sheep.

Once Lora saw Miss Farr playing a mouth organ, and she would not let her mistress have any peace until the mouth organ was put between her teeth. Lora blew in and out, and it was just mouth organ noise, but the seal loved it no end. She would not stop and it was quite a job to take the mouth organ away from her. But the instrument that had the strongest effect on Lora was a bamboo pipe. Even when Lora was outside the house, playing with the otters and enjoying herself, she would come in as soon as she heard the sound of the pipe and sit still, as if she were made of stone. Lora never tried to sing with the pipe, she just listened, and her eyes seemed to look far away, as if she were in a dream. It was as if the sound of a bamboo pipe cast a spell on her. It would be interesting to find out if all wild seals like to listen to the music of bamboo pipes.

To tell you one more story about Lora and music: people had heard of Miss Farr's musical seal and so it came that when there was a ceilidh — a gathering for music, dance and song — in Aberdeen, Miss Farr was invited to come with Lora. The people who sent the invitation thought that, after some of the singers and players had made their contribution to the musical evening, Lora would wail her "ba ba" and it would be good fun. Well, Miss Farr made the car journey to Aberdeen with Lora and brought her to the ceilidh. The first to sing was a lady, but she had hardly sung her first note when Lora started her own famous wail. The audience was roaring with laughter, but poor Miss Farr was deeply ashamed of her pet. She wanted to put Lora outside, or in the kitchen, but Lora made such a racket, barking and howling, that she had to be brought back to the hall where the ceilidh was held. And once she was there, all the performers, singers, accordion players, violin players, had to give up — Lora had to sing her song and nothing could stop her. So in the end, the whole musical evening was just one performance, and that was Lora's.

Lora stayed with Miss Farr for seven years. Then one day Lora went for a swim in the loch — and never came back; no one knows what happened to her. But Miss Farr will never forget her pet seal, nor will anybody who reads her story in the book *Seal Morning*.

4

The Snail

We have now heard about a trunk-animal, the seal. Now we shall look at a creature that is again a head-animal, like the cuttlefish, yet it is quite different. Let's first think about the head again. Our head not only has eyes, ears, nose and mouth, but above these the head has the round part of the skull made of very hard bones. Inside the hard bones of the skull there is the soft brain. The brain does not just lie inside the skull, but it floats, because around the brain there is water. Not a lot of it, but just enough to keep the brain floating.

There are creatures which are rather like this. Something hard outside (like our skull bones), something soft inside, and this soft inside is floating in a watery liquid. Such creatures are the shellfish, like oysters and others. But we shall look at a creature that is not a shellfish, but a relative, a kind of cousin of the shellfish, and that is the snail. The snail is also a head-animal, but an animal that is more like the upper part of the head, the skull and the brain. The snail's house is like the skull and the soft snail is like the brain in the skull. Snails are, of course, not as beautiful as some of their cousins in the sea. The seashells often have quite wonderful shapes and colours, especially those found on the shores of hot countries like Africa. Our ordinary snail does not have as beautiful a shell as these sea creatures but it has other wonders.

You could really say, a snail has eight wonders, or eight secrets. The first secret is of course the shell or house of the snail. This shell is really a kind of house, a house that the snail itself has built. The snail can get right inside the house if it wants to, but even when it is moving about it has to carry the house on its back, like a rucksack, because the snail is firmly attached to its house.

But the real secret of the snail's house is the way it is built. We can only discover this secret if you take an old shell where there is no longer a snail inside and then cut it in half with a fine saw. And then we find that the inside of the snail's house is built like a spiral staircase that gets narrower and narrower towards the top. This is the secret. The snail starts when it is little with the tiny spiral on top and as it grows, the spiral becomes wider and wider. No human builder could make the spirals more exact than the snail does. There is a rock called lime, and we use limestone for making cement for building. The snail also builds its house from lime, and so snails can only live where there is at least some lime in the soil. The snail eats the lime and then "sweats" it out where another bit of the spiral house is needed. If there is no lime in the soil, snails sometimes crawl up houses with limestone walls and eat the lime they need there. So the snail's house, built from lime in a very regular spiral pattern, is the first secret.

The second secret is the way the snail moves. The snail has no legs or feet, the whole body is its foot. But you might just as well say that the body is rather like a tongue, and the snail crawls on its tongue. It is difficult to find out how the snail crawls if you only see it in the garden, but if you put the snail on a glass plate and look at it from below, you can see that it makes ripples on the sole of its foot and with these ripples it moves along like a caterpillar. But this is not all. The snail oozes out some shiny liquid on which it ripples along, it is this liquid, called mucus, that leaves a shiny trail behind. So the snail does not really tread on earth, but glides or ripples on the liquid. Our brain does not ripple and move but it is also carried by liquid. This movement, rippling on liquid, that is the second secret of the snail.

The third secret of the snail has again to do with its house. You see, we humans can come and go from our houses, we go home for shelter and for rest, then next day we are ready to go out into the world again. Now the snail also uses its house for protection, if some danger comes along it quickly disappears right inside the shell. But, like us, the snail must also go out into the world to find food and to travel to new places, and so it too must leave its house — but not completely. It is mostly at night that the snail gets going, and it can be very agile. It can soon climb to the top of a nice juicy plant, or it can even make its way right up a tall tree, but it always carries its house with it, it never leaves its shell completely. Even when it is very busy with other things — travelling, climbing, eating, meeting other snails — there is never any danger that it might accidentally drop or lose its shell. And the secret is this: on the inside, the snail always has one tough muscle of its body firmly attached to the shell. This tough muscle of the snail is also close to the part where the snail's food is digested and the whole thing is coiled up in a spiral, just like the shell itself. So we could say that although most of the snail can come out of the shell, its stomach is always inside and it is close to this part that is firmly attached to its house — no matter what.

★

We have heard three of the snail's secrets. The first was the secret of the house: that it is built like a staircase in a beautiful spiral. The second secret was the movement of the snail: it swims, as it were, it makes ripples on its belly and oozes out a liquid, and on that liquid it swims. That is the reason why on very hot and dry days snails don't move about, they stay in the shade, otherwise they would use up all the watery liquid in their bodies and not find any water to drink. And the third secret is that the snail has a part of its body grown together with the house, and that part is near the stomach.

The next secret of the snail is its eyes. The eyes are not in hollows in the head, they are at the end of two long horns or stalks, which are of course not horns, they are soft feelers. The snail has its eyes at the end of these long things. And so it can swirl its eyes round and even see what is behind it. But there is something else about these eyes. Just like the finger of a glove can be turned inside out; the snail can do this with its eyes. If you touch a snail the horns are drawn in, rather like pushing the finger of a glove inwards.

The fifth secret is something one would hardly expect of a

wet, cool creature like the snail. The snail likes warmth — as long as it is not too dry — it needs warmth as much as it needs rain, water. So what does the snail do in winter? The snail looks for a place where it is well hidden, then it withdraws right into its shell or house and then out of mucus it makes a door, or rather a lid, that dries and covers the big opening. Yes, the snail can close its house completely, so completely that the cold outside cannot reach it. And so the snail sleeps warm and comfortably in its little house right through the cold months of the year.

The snail must breathe, it needs air to breathe just as we do. But the snail's breathing is not like ours. Snails have no nose to breathe with, nor do they breathe through the mouth. That is the sixth secret. The snails have a special hole, just where the shell begins, a little inside the shell, it has a special breathing hole. And the breathing hole does not draw air in and out as we do, it is just a hole and air comes in and out on its own. If the snail were to move very fast it would not get enough air in this way — you know how we puff when we run — but as snails never move very fast the little air that goes in and out is enough for them.

So we have now the secrets of the spiral house, of movement, of attachment to the shell, of the eyes, warmth (a door that closes the shell in winter), and the secret of breathing, that's six. The next secret, the seventh, is the snail's tongue. The snail's tongue is really a little rasp or file. And when a snail eats a leaf, the tongue rasps or files the leaf into tiny bits.

The last secret, the eighth, is that the snail lays eggs. The snail does not lay its eggs just anywhere, because some birds, like the thrushes, would eat them up quickly. The eggs are quite small and they do not have a hard shell like hen's eggs; they only have a fairly thick skin. And to protect the eggs, the snail digs a hole about 15 cm (6 in) deep and lays its eggs in the hole. But the snail is a soft creature, it digs with the hind end of the soft belly. It takes hours to scoop and bore a 15 cm deep hole with the soft hind end. Then the snail lays about sixty eggs into the hole, and closes it with earth. But then the eggs are left to themselves; the snail does not bother about them any more. And after a month the little snails come out and slowly build their spiral houses and can look after themselves.

If one snail lays sixty eggs, and every little snail could in time lay another sixty, and if this went on, the earth would soon be covered with snails. But the wisdom that works in nature has seen to it that there won't be too many snails. The birds — thrushes and starlings and especially blackbirds — eat snails. Blackbirds take a snail with its shell and the snail of course withdraws, goes inside the house. But the blackbird knocks the shell against a stone, breaks the shell, and then eats the snail inside. It is a pity, but otherwise there would be too many snails, they would eat every green leaf there is. And so it is a good thing that the blackbirds and thrushes eat snails and so see to it that there are not too many of them. It is the great wisdom of nature, the wisdom that God put into nature, which sees to it that there are not too many of any kind of animal.

We have now heard about the eight secrets of the snail. The first secret was the house. There must be some wonderful cleverness of the snail to make these lovely spirals. But the snail uses up all the cleverness it has to make the spirals, so it has no cleverness left for anything else, and that is why you could not train or teach a snail as Miss Farr taught things to her seal. The snail has only that one cleverness, and can't use it for anything else.

And it is the same with the cousins of the snail, the shellfish in the sea. There is, for instance, the oyster. The oyster gives a beautiful shiny, pearly colour to the inside of the shell; it is called mother of pearl. No human artist could match anything so lovely, but the oyster too uses up all its cleverness to make these lovely colours and it has not anything left for anything else. Perhaps it is just as well. Perhaps the oysters would be much cleverer than we are if they did not use it all to make the pearly inside of their shell. And the snail would be much wiser than we are if it did not use all its wisdom to build beautiful spirals inside its house.

The Harvest Mouse

We have now heard about the cuttlefish, which is a head-animal, the seal, which is a trunk-animal, and then the snail with its secrets was another head-animal. Now I want to tell you about another trunk-animal, but one that is very different from a seal. Seals are quite large creatures — as we remember, four weeks after they are born they weigh twice as much as a ten-year-old child. But the animal I want to look at now is so small and so light it weighs less than a letter, about 5 or 6 grams ($^1/_5$ oz). And as to size, the animal is only 6 cm ($2^1/_2$ in) long. Yet it is not an insect or a little bird, but a mammal, like dogs or cats. It is almost the smallest rodent in Britain (only the pygmy shrew is smaller). And it is a trunk-animal, because when you look at it you can see that the main part of it is really the trunk, the head is just a continuation of the trunk and the limbs are very tiny.

This little creature, 6 cm long, weighing only 5 or 6 grams, is a particular kind of mouse called the harvest mouse. The harvest mouse is rather rare in Scotland, but there are plenty in England. Imagine at harvest time walking past a cornfield in the south of England, there is a little breeze blowing, and ripe wheat sways in the wind. Being an observant person, you notice that one stalk swings much more than the others. So you go a bit closer to investigate, and what do you see? A little furry bundle has climbed up on the stalk and is just about to bite off the top part with the ripe ear of wheat on it. This little furry bundle is a harvest mouse.

It does not stay to let you inspect it for long, but even with a short glance you can observe quite a bit. The harvest mouse has brownish-red fur, sometimes almost orange, but the underside of the body is white. Its ears are short, much shorter than the ears of the common mouse in houses. The tail is quite long, as long as the body, but it has no hairs. But the funny thing is that the harvest mouse uses this tail as an extra arm or leg for climbing up on a corn stalk. It curls the tip of the tail around the stalk as it climbs up, just as some monkeys do. If you see a family of harvest mice at play in a cornfield, it is really like monkeys playing high up in the trees. Of course, for a tiny harvest mouse a stalk really is as high as a tall tree and a cornfield is a mighty forest. And when the harvest mice are at play they climb, swinging from one corn stalk to another, even leaping from one stalk to another. Because the harvest mouse uses its tail a lot, it looks after its tail, and keeps it clean. So you can sometimes see one of these tiny mice sitting on the ear of corn, squatting on its hind legs and holding its tail with its forepaws while it diligently cleans it with tongue and paws, all while balancing on the top of a corn stalk.

The harvest mouse is really the most graceful and entertaining of all the creatures of the fields, it is a delightful animal to watch, although it is not easy to watch it for long as it has many enemies and flees as soon as anything large gets near it.

They make lovely pets, but only if you give them a proper home and keep several of them together, for the little harvest mouse is a sociable creature, it likes company and is unhappy if there are no others about. The best place to keep them is a large box with glass front and sides so that the animals can have plenty of light. On the bottom you put a good layer of sand and stick some corn stalks into it — soon you will see them perform acrobatics that are just as exciting as anything you can see in a circus. They are lovely little creatures, full of life, quick, not at all like the slow snails, and they are soon tame and used to coming to you for food.

There is one thing in which the harvest mouse shows a quite surprising cleverness, in one way it is so clever that you would hardly believe it of such a tiny creature with a very tiny brain.

★

The harvest mouse is a wonderful acrobat, performing amazing feats of climbing, swinging, leaping, among the corn stalks, but even more amazing is that the harvest mouse builds a nest. It does not lay eggs like a bird, it gives birth to little ones as dogs and cats do, yet it builds a nest. And its nest is as wonderful, or even more wonderful than any bird's nest.

The nest is made of blades of grass and blades of wheat. And these blades of grass and wheat are so cleverly woven and plaited together that they form an almost perfect globe, a ball about 8–10 cm (3–4 in) wide. And the weaving is done so well that you could take the hollow little ball made of grass and roll it on the floor and it would not fall apart. It is very strong, compact weaving; the mouse takes stalks with their leaves, uses the stalks as scaffolding for strength and interweaves the leaves. For one nest about a hundred stalks are used. This little sphere or ball is hung up between two or three corn stalks about 30 cm (1 ft) above the ground. In its own way this nest, built by the tiny mouse, is as marvellous as the snail's shell.

This nest is really a nursery for the little ones for the first two or three weeks after they are born. There are usually five or six of them and the mother suckles them, lets them drink her milk inside this nest. Inside the nest there is even soft bedding for them because the mother has taken the leaves of grass and split them up into smaller bits that make a nice soft heap at the bottom of the nest. When you see such a harvest mouse nest for the first time, you may be puzzled, because nowhere can you see an entrance; it is closed all around. How does the mother mouse go in and out? She pushes or bends apart two stalks that are woven into the nest and squeezes in between them. And once she is through, the elasticity of the stalks closes the gap again, and you can't see any opening. Is it not wonderful that this little creature can build such a round, compact nest? If we tried to make a nest from blades of grass and stalks of grass and corn by weaving them together, it would hardly hold together. But the little harvest mouse builds these cosy little spheres or balls as a nursery for its young.

The harvest mouse not only builds a nest for its young, it also builds another nest for itself for the winter. The mouse does not take as much trouble with the winter nest as it does with the nursery nest for the young. The winter nest is built much looser, but it is lined inside with moss, and this keeps the harvest mouse warm and comfortable in the cold season. The harvest mouse also uses the winter nest as a store or larder. It puts in a store of grains and grass seeds to feed on when the fields are bare. Harvest mice don't hibernate — they don't sleep in winter — but they do sleep a lot and they don't move about much in the cold season. If harvest mice find a haystack they see a big nest all nice and ready, and quite a number of them make the haystack their winter quarters. The farmers don't like it very much, as the mice eat some of the hay that is needed for the cows. But the harvest mice do not do a lot of damage and so one should not begrudge them the little they take. Moreover, the harvest mouse also does some service to the farmer: in spring and summer, before the corn is ripe, the harvest mouse feeds on insects, which can do more damage than the mouse.

The Red Deer

After a tiny trunk-animal, the harvest mouse, I want to tell you about a much larger trunk-animal, a trunk-animal that is at home in Scotland. That animal is the red deer. You may wonder why I call the deer a trunk-animal, after all, you can quite clearly see the head on the graceful neck with its antlers, and also the slim but strong legs. But deer use their heads as a kind of extra limb, to pick up food, to push aside branches, and sometimes to fight each other. They use their heads as we use our arms and hands. The head is used as a kind of strong arm in the service of the trunk. The head is a servant of the trunk. And so are the legs: forelegs as well as hindlegs carry the trunk. So, both head and limbs are servants of the trunk and the trunk is the most important part of the whole body of the deer.

The red deer is not red during the whole year. In autumn and winter and until April, the fur is greyish-brown, long and rough. In spring the animal sheds the winter coat and grows the reddish-brown hair which gives it its name. Around the tail there is a light patch. As the male deer are called stags, the females are called hinds, and the young ones are called fawns or calves. It is only the stags that grow antlers, the hinds haven't any. There is a great difference between the horns of a bull and the antlers of a stag. The horns of a bull grow once, and once they are grown they stay on his head as long as he lives. But it is not so with the antlers of a stag. The stag sheds his antlers every year between February and April — they just come off. It has been found that if a stag loses his first antlers on a certain day, say the fifth of April, he will always lose his antlers on the fifth of April in the following years.

But soon after the antlers of last year have come off, new ones begin to grow, and the new antlers have more spikes than the old ones. In the first year there are only two straight horns, in the second year each horn branches into two spikes; it is called "four points" and so it goes on. A stag with 12 points or more is called "royal," and some stags have lived long enough to grow 20, 30, and even 40 points, but this is rare. A 14 pointer already has a very impressive crown of antlers. After stags throw one pair of antlers off they start growing a bigger pair that is full size in four months. It is like a plant that dies but grows again from a bulb.

Deer are really forest animals, they like to be among trees. You might think that antlers, especially when they have grown big, are in the way when a stag wants to run in the forest. But this is not so. You can see stags running at full speed in forests where the trees are very close. He throws his head well back so that the antlers rest on his shoulders and so any small branches in the way are easily brushed aside or broken off.

In the autumn the stags look for hinds. You can hear them roaring, and if two stags like the same hind — which happens very often — then there is a fight. It is then that heads and antlers are used as weapons. These fights between two strong stags are wild and terrible, but it only happens very rarely that a stag is killed in such a fight. In most cases one stag just has to show that he is the stronger one, and that is enough.

In one such fight — between a quite old stag, an 18 pointer, and a younger royal with 14 points — the two stags retreated for a distance of about 20 m (65 ft) and then charged at full gallop at one another. When they clashed it was with such force that sometimes they were both thrown back on their haunches by the force of the impact. Yet they got up and charged again, stabbing each other with their antlers. At last, after half an hour, the younger stag managed to strike with such force at the ribs of the older one that he was knocked over and fell on his side. As he struggled to get on his feet the young one gave him another hard blow, and then the old one had enough; he leapt up and fled as fast as he could. But the victorious younger stag was so exhausted that he sank to the ground with heaving flanks and did not move for half an hour until he had recovered his strength.

At the time of the year when the fighting spirit is on them, stags have even attacked people. But one thing you can see from the story of this fight: the stags use their heads and antlers like an arm with a weapon in it, and so their heads do not, as ours do, rest quietly. Their heads are a kind of limb.

★

Let us look at the feet, at the hooves of the red deer. They are, of course, not the only animal that has hooves; so have cattle, goats, sheep, pigs and horses. Horses have a single big hoof on each foot; the others have split or cloven hooves. If you ask a little child, "How tall are you?" they will try to stand on their toes and, perhaps, even lift their arms to make themselves taller. Now, with all the animals that have hooves and horns something similar happened a long, long time ago. They grew horns and antlers, like the child lifting his arms, and they got up on their toes. The hooves are really only the toes of their feet, and the toenail has grown around so as to give it more strength. Deer — as well as cows, sheep, goats and horses — walk on their toes all the time. It is as if they had once tried to become upright, like human beings, but could not. In each hoof of the deer there are two toes, another toe is only a little stump at the back of the foot. Just as we, when we run fast, run more on our toes than on our heels, so deer, being on their toes, can run very fast.

Deer are animals of forests and woods, they like to be among trees. They feed on acorns and horse chestnuts, but they also like something else which is not good for the forest. They like to nibble the bark of young trees, and if a young tree loses its bark, it dies. And so if there are too many deer in the wood they destroy all the young trees. In time the old trees fall and there is no more forest, for there are no young trees to take the place of the old ones. And that is why it is necessary to hunt deer, to shoot some of them; otherwise we would not have any woods or forests left. In the old days there were wolves that killed many deer and so kept their numbers down, but the wolves were killed by humans a long time ago and so we must kill some deer every year to keep their numbers down.

Many forests in this country have a keeper who does some of the hunting and shooting of the deer that must be done. Following the deer in a hunt is called deerstalking. It is not easy to stalk a stag because you must get near enough to be certain to kill him with the first shot. It would be cruel only to wound and hurt him. Once a keeper was stalking a stag. It was a fine beast with twelve points, feeding on an open stretch, so there was no tree which the hunter could use for cover. He went down on hands and knees, crawling towards a large boulder. If he could get to the boulder, he would be near enough to chance a shot.

But as he crawled forward he disturbed a large bird, a grouse, and it flew up. Usually the sudden flight of a grouse is a warning for a stag and he races away. But this time the stag took no notice of the bird and continued to feed. The keeper gave a sigh of relief, he thought now he would get the stag. Then something quite unexpected happened. A golden eagle saw the grouse, and came swooping down, caught it and flew up onto a grassy knoll near the grazing stag. All this commotion still did not disturb the stag. In the meantime the hunter was at the boulder and lifted his head and got his rifle ready to shoot. But just then the eagle decided that the stag was disturbing *it* and it flew with flapping wings onto the stag's back, riding him like a rider on a horse. This was too much for the stag. Off he went as fast as he could and, of course, the gamekeeper had lost his chance to shoot the stag. When the eagle had driven the stag some distance away, it left him and flew back to eat its grouse. But the stag probably never knew that this ill-tempered eagle had saved his life.

Corrie, the little deer

John Cameron was gamekeeper on one of the big estates, which were mostly woodland, where red deer roamed. He was out stalking deer when he passed through a corrie, that is a hollow in a hillside. And there — he nearly stepped on it — was a little red deer calf lying in the bracken. It was so little that it could not be more than a few hours old, and it lay flat on the ground without moving.

"Well," said John Cameron, "you wee thing, I wonder what made your mother leave you like this. But I know a lassie and a lad who will love to have you as a pet and to look after you." And he picked the calf up and took it in his arms and went down the mountainside to his home.

His two children, Ivy who was twelve and Joey who was ten, danced with joy when they saw what their father had brought home for them. They decided that an old summer-house across the lawn could be the calf's home, and they called him Corrie because he had been found in a corrie. Joey brought armfuls of straw, Ivy prepared the first feed: milk mixed with some chalk and with white of egg in a feeding bottle. At first it was not easy to make Corrie suck at the bottle, but after a little while he learned the trick and drained the bottle to the last drop.

That night the children did not sleep very well, they were so excited about their new pet. But even if they had slept well the noises that went on outside might have awakened them, for there was a sound like barking and grunting going on just outside their windows. Now children who grow up near forests are used to all kinds of animal noises at night and so neither of them got up to find out what was making the noise.

Ivy got up early next morning to feed her pet. When she opened the front door, the feeding bottle in her hand, she saw a fully-grown deer, a hind, standing by the summer-house; it must be the little one's mother. Ivy could not even imagine how the hind could have found her way to the summer-house, but when the hind saw the girl she leaped away and disappeared up the hills. The hind never came again but even John Cameron, who knew a lot about deer, could not explain how the hind had come all the way and had jumped the deer fence on the way.

As it was Ivy who fed Corrie it was only natural that the little stag came to look upon her as his mother. He was quite tame and friendly with all human beings, but he loved Ivy more than anybody else. And whenever she was standing or sitting by herself Corrie would edge himself close against her and raise his head to be stroked.

It was a warm summer and Ivy and Joey would often go down to the loch for a swim, and Corrie came with them and swam with them. He was even a better swimmer than they were and could swim out farther than they did. The children had been warned by their parents to do their swimming only in one part of the loch where the water was not too deep and there were no dangerous currents, but one day Joey, the boy, saw a white motor-launch that was moored further out in the bay. Joey was curious to find out more about the motor-launch; it did not look as if it was far away and so he decided to swim out to it.

However, the boat was really further out than he had thought — as he discovered while swimming — and Joey began to feel his muscles getting tired. Now further out the water was much colder, and when at last he got to the launch, he was so exhausted that he had not even the strength to climb aboard. He could only hold on to the mooring rope, but even that he could not go on doing indefinitely. He was getting colder and colder, he was shivering, and he knew he could not last much longer. And then — there was Corrie, swimming gleefully around him, for Corrie it was just a game. But Joey quickly let go of the rope and got hold of the long hairs on Corrie's hindquarters and so Corrie towed him across to the safety of the shore. Joey, being an honest boy, told his parents what had happened and he had a good row for disobeying them. But now the whole family knew that Corrie had saved his life.

The little stag calf had saved Joey's life and you can imagine that both children, Joey and Ivy, now cherished their pet even more than before. But Corrie grew — and he grew stronger. There came a day when Joey had some work to do and Corrie came and bothered him and would not go away, until Joey became impatient and gave Corrie a sharp smack on the head. Corrie stepped back, lowered his head and charged. Joey was caught unawares and he was sent sprawling into a bush. And as soon as he was on his feet again Corrie came on a second time and threw him over, whereupon the little stag trotted off, apparently very pleased with himself.

Unfortunately, from that day Corrie had got it into his head that it was great fun to send people sprawling. When a strange child came to the house, that was the welcome it had from Corrie, and some younger children were badly frightened. And so in the end Mother and Father Cameron said, "Sorry, we can't keep Corrie any longer, he must go, before somebody gets seriously injured."

The two children were not happy, but they realized that it was necessary to part from Corrie. Corrie was taken by car to a wild part of the mountain that was enclosed on all sides by a deer fence. Ivy put a leather collar with a little bell on it around Corrie's neck and she said, "I hope this will make sure that none of the gamekeepers will ever shoot you." And then they all left, leaving Corrie behind, and Ivy cried. Corrie was not at all happy in the mountains. He was used to human company; he did not care for the red deer. And so, after a few nights, he somehow got out of the deer fence and onto the open road.

As Corrie was trotting along the road, a motor car came from the opposite direction. The driver stopped, and Corrie stopped. Corrie thought the motor car should get out of the way, the driver thought the stag should go, and so they were both stuck. Soon other cars came behind the first one, there was a whole queue held up, until one man managed to chase Corrie off the road. But Corrie was still determined to get back to his young mistress. The next day he knocked an old lady off her bicycle, and in the evening he knocked down some children coming back from school. But Corrie did not manage to get back to the Cameron family. Some young men — who had heard of the young stag that was making the roads unsafe — chased Corrie into a big box that was then closed. And they took Corrie and sold him for a few pounds to a game park in England.

You would think that Corrie was well off in a large park where there were many other deer, where there was plenty of food, where he could roam quite freely. But this lovely wooded park in England was not at all like the Highlands of Scotland and Corrie was a Highland deer. And because Corrie was not happy there, he became the least tame, the wildest of the deer in the park. In the end the gamekeeper of the park said, "I don't want this wild fellow in my herd." And so Corrie was again driven into a big box and brought by car to an estate where the owners wanted to have a proper stag hunt with horses and hounds as it used be in olden times.

So on this estate in England Corrie was set free and after a time the huntsmen on their horses and the barking dogs came after him. The hunters expected that Corrie would try to escape towards the game park that had sent him, but they were mistaken. Corrie, by now a strong grown-up stag, had only one desire: to get back north, to the Highlands. And with his strong legs and his cunning, he escaped from the hunters, he got away from the estate. Corrie had no map, no compass, he could read no signposts, he could not ask anyone to show him the way, yet he knew which way to go. It was a journey of hundreds of miles — he avoided cities and villages, he moved mostly by night over fields and farms, often pursued by dogs. But he got away every time, and for the last part of the journey he had to swim across a large bay. One evening some fishermen, to their surprise, saw a stag leaping into the sea and swimming across the bay. They all wondered at this strange sight. Yet, some weeks later the gamekeepers in the mountains where Corrie had been born saw a stag with a leather collar and a bell on his neck. And so it became known that Corrie had returned to the Highlands to which he belonged.

The Hedgehog

Now we come to an animal that is very different from the deer, the deer with its graceful neck, long slender feet and soft fur. Our animal is rather small, the feet are so short that one can hardly see them, and instead of beautiful soft fur this animal bristles with sharp spines. It is the hedgehog. If you have once seen a hedgehog you know that what you have seen was mainly trunk, the head is just a little snout peeping from under the bristles, the legs are short. And so what you see as a bristling ball is almost all trunk; the hedgehog too is a trunk-animal.

The hedgehog is a wonderful example of the way in which nature protects her children. The deer have antlers and they can run fast from any danger, the snail can withdraw into its house, the harvest mouse has its nest, the hare is a fast runner. But the hedgehog has its prickly armour and this armour really protects the hedgehog so well that it does not matter that he cannot run as fast as some other animals, or burrow into the earth.

When the hedgehog is undisturbed, the spines do not bristle, they lie back. But as soon as there is the slightest indication of danger, the hedgehog rolls itself into a tight little ball and every spine stands up. And this standing up of the spines is so fast that your eyes can hardly follow it. One moment the spines lie flat — and the next they are up. But the hedgehog can make itself into a little ball without making the spines stand up, and it does this when it has to go downhill on a steep slope. It then becomes a little ball and just lets itself roll down. The hedgehogs that are kept by people as pets will do this rolling for fun if they are put on a wooden board that is sloping down. Hedgehogs will even let themselves drop from a height of 1.5 m (5 ft) and they fall so skilfully that the spines act like a sort of cushion.

Hedgehogs are very useful animals for the farmer: they eat insects, slugs, and all kinds of little pests that could ruin the harvest. Hedgehogs are animals which hibernate, that means they sleep through the cold winter months. They just make a little hollow for themselves, cover themselves with dead leaves and weeds and roll themselves into a little ball. In the rains the weeds and leaves are flattened down and so, after a time, you can't see anything to show you that here a hedgehog is sleeping.

Hedgehog mothers have about three to five young, and the baby hedgehogs look quite strange, for their spines are still soft and pinky white. It takes nearly a month before the spines are hard and then the young ones leave the cosy nest the parents made for them. The nest is just a hole in the ground, made comfortable with dry grass and moss.

One of the most interesting things about hedgehogs is that they fight venomous snakes. When the hedgehog meets a snake, it first moves slowly, with lowered head, until it is close enough to dart with lightning speed and bite the snake. The next moment it has rolled itself into a ball, with spines facing in all directions. The snake is angry and darts its head at the spikes, but it only gets hurt. Again the hedgehog unrolls, bites and rolls up again before the snake can get at it. The snake is even more furious and attacks the spikes, but again it only hurts itself, and after a few times of this, the snake is so hurt and has lost so much blood, that it has not much strength left. And then the hedgehog bites at the back of the snake's head, and that is the end of the snake. So you see, the hedgehog can be quite a dangerous enemy for someone who, like the snake, loses its temper and forgets the prickles.

Prickels, the hedgehog

Here is a story of a hedgehog who made friends with a boy. The boy's name was Tommy and he was the son of a gardener, a man who grew vegetables and flowers for market. Tom always helped his father and one of his jobs was to go in the evening and put coal into the boilers which heated the greenhouse. One evening he had just finished stoking the boilers and was walking back to the house when he heard a terrible scream, something like the scream of a rabbit, yet also more like the call of a cat. The scream came from a corner of an old stone wall and Tommy went to see what kind of creature had made this heart-breaking noise. A long time ago his father had put a trap for rats by that wall, but no rats had been caught and everybody had forgotten the trap was there. Now Tommy saw not a rat but a little hedgehog with his foot caught in the trap and screaming his head off. Hedgehogs have voices, and very loud ones if they are in pain. Tommy opened the spring of the trap and saw that the hedgehog's little paw was badly damaged.

Tommy was a kind boy, so he took the hedgehog home and bathed and bandaged the wounded leg. There was an empty rabbit hutch near the house and that is where Tommy took Prickles, as he called his pet, putting him on a nice bedding of straw and feeding him with bread and milk. The wounded paw healed in less than a week and with the healing Prickles lost his fear of human beings. At first he always rolled up, the quills bristling in every direction so that one could not touch him, but soon Prickles took food from Tommy's hand and would lie there with his prickles quite flat when the boy touched him.

One day Tommy even discovered a curious trick that his strange pet could do. Tommy wanted to have a look at the wounded leg and so he put Prickles on his back and with a straw touched the hedgehog's hind feet just inside the upper part. And when this happened Prickles gave a loud squeak, and went on squeaking as long as the straw tickled him. Yet he did not seem to mind, and as he let it happen without trying to get away or to roll up, he even seemed to like it. So whenever one of Tommy's friends came, Tommy would say, "My hedgehog can sing. I will show you." And then he tickled Prickles' hind legs with a straw and Prickles would squeak loudly and happily.

But Tommy's mother was getting rather tired of providing bread and milk for an animal that could well fend for itself, and so Prickles was put back into the garden. Daybreak and evening are the hedgehogs' hunting times and during that summer Tommy saw his little friend quite often, sometimes in the early morning, sometimes in the evening, trotting about. The animal's legs being so short, it looked as if Prickles was slowly rolling along on wheels, but sometimes Prickles could put on an amazing speed, as he showed in his encounter with the neighbour's cat.

The neighbour's cat was a large fellow who enjoyed hunting birds and rabbits and who had even fought and killed weasels. That evening the cat was sitting on the stone wall when Prickles appeared and wandered about among the rose bushes. The cat flattened out, and Tommy watched, wondering what was going to happen. Prickles, apparently not noticing anything amiss, came nearer and nearer to the wall. The cat must have thought Prickles was a rabbit, for she waited until Prickles was directly below, and then just dropped on him. But in the half-second it took her to drop from the wall, Prickles had rolled up, his head down and his spikes pointing in all directions. The cat fell onto these spikes with all the force that was meant to kill a rabbit. The cat gave one loud screech and shot off, and I don't think that she would ever make the same mistake again. Prickles too was much upset by this encounter and fled with great speed under a peony plant.

It was soon after this that Prickles discovered that the boiler house was a wonderful hunting ground for him, for it was infested with cockroaches as well as mice. Prickles killed some of the mice but all the others went away, for mice cannot stand the way a hedgehog smells to them. And the cockroaches too soon became fewer and fewer; so Prickles made himself rather useful to the gardener.

★

When autumn came, Prickles prepared for his hibernation, and Tommy had the opportunity of watching him. The gardener had swept up a heap of fallen leaves at the bottom of a broken-down wall. Now Prickles mounted the wall and when he was on top he just dropped into the heap and, of course, a lot of leaves got stuck in his spines. And again Prickles climbed up the wall and let himself fall down on the dead leaves, and got more of them stuck to him. And he carried on until all his prickles were just a mass of leaves from his skin outwards. It was a thick cover of leaves which would not only keep him warm but would also keep out any rain or damp.

But after this burst of activity, Prickles became very slow and listless, as if he was in a daze. Before that time he had always been quick enough to get out of the way of any pair of feet that came too near to him. But now people had to take care not to step on him, he was that slow. And then one day Prickles disappeared for a while; he was not seen for a whole week. Then Tommy saw, in the corner of an old shed, a big ball of dry grass. He was curious; what could it be? So he unravelled the ball of grass — and there was Prickles curled up snugly for his winter sleep. Tommy was sorry that he had disturbed his friend, so he put the half-opened ball back into the corner, but Prickles was no longer interested in that grass, he had been disturbed, and when Tommy looked the next day, Prickles had gone.

Some more weeks passed, it was getting cold and Tommy's father was cutting away parts of a creeper that had grown along one wall of the house. The creeper was densely matted with leaves and the gardener was clearing the leaves away when he discovered Mr Prickles, six feet from the ground, snugly packed away on a platform of leaves. But with the cutting of the creeper Prickles lost his second hidey-hole and disappeared again. A few days later Tommy's father, doing some work, got very warm, took off his jacket and put it on the grass by a bush. Later he forgot about the jacket and went the next morning to fetch it. But in the meantime the jacket had acquired a tenant — rolled up in one sleeve was Prickles. Of course poor Prickles was disturbed again, and disappeared.

Some days later Tommy's father went into a shed where he

kept large bundles of strong dry grass that is used for tying plants and flowers. He lifted a bundle, and there was Prickles, so he was turned out again. The house had a garage and, as it was a large garage, one corner of it was often used to store things out of the way and among these things was a thick rug that was rolled up and left there. But as the days became cold, very cold, Tommy's mother remembered the rug and thought it would be useful. She went to fetch it, and as it was unrolled — there was Prickles. So once again his sleep was disturbed and he had to look for another safe place.

For a time it seemed that at long last Prickles had found a place where he was not a nuisance to people, for he was not seen or found. But one evening Tommy was in the boiler house shovelling coke into the furnace when he suddenly saw a prickly ball among the coke on his shovel. Poor Prickles had thought that among the coke in the warm boiler house he would be safe.

"Well," thought Tommy, "I must find a place for him, the poor fellow got into trouble every time he picked his own place." And he took Prickles to an old rabbit hutch and packed it tightly with straw. And there Prickles stayed at last for the rest of the winter and slept without any further disturbance until spring came and he could again move about in the light and warmth of the sun.

The Eagle

We come to a kind of animal we have not had before: a bird — the golden eagle. Before we look at the golden eagle I want to say something about birds in general. The great and wonderful thing about birds is that they can fly. If we watch birds in flight there is something so beautiful about it that the way animals move on land, or even our walking, seems quite clumsy. It is a lovely thing, the flight of birds, and it is so because the birds are built by the wisdom of nature to fly in the air.

If we look at the shape of a bird, we might remember the seal who has a shape to cut through water at great speed; the seal has a streamlined shape. And in the same way the shape of a bird is streamlined for flying or gliding through the air. But because the body is streamlined, it shows also that the birds are trunk-animals, the trunk is the main part. You can already see from outside, just looking at a bird, that its body is built for moving through air, it is streamlined. But there are things you can't see from outside, there are secrets of the bird (as there were secrets of the snail), and these secrets make it possible for the bird to fly so beautifully.

The first secret is in the bones of a bird. The bones of a bird are not only much thinner than the bones of land animals, they are also hollow, they are hollow tubes. The large bones in a bird's wings and legs are hollow; they are not filled with bone marrow as ours are. This makes a bird so much lighter. These thin, hollow tubes as bones are the bird's first secret.

The bird's second secret is its lungs. The bird has its lungs in its breast as we have, but it has something that we do not have. From the lungs of the bird long sacs go into its neck, chest and belly, long spaces that are filled with air. They are called air-sacs. So when a bird breathes in, the air does not go only into its lungs, it goes into all the air-sacs, into all parts of the body. So the whole body of a bird, not only its lungs are filled with air. That's why it's so easy for a bird to be carried by air, to be air-borne, just as we feel carried by water when we float. The bird has air-sacs that go from the lungs into the whole body, and because the bird has so much extra air in it, it never gets out of breath. If we run fast for too long we start puffing, we get out of breath, even stags or horses or dogs get out of breath, but never a bird. Of course, a bird can get tired, but never out of breath, because it has these wonderful air-sacs and so can take in much more air (for its size) then we can. The air-sacs are the second secret of the bird.

The third secret is the wings. The wings are nothing else but arms and hands that have been changed into instruments for flying. Where we have arms and hands, where the animals have forelegs or forepaws, the birds have wings, and when klthey spread their wings, it is as if we spread the five fingers of our hands. It is wonderful for the bird that its forelimbs have become wings that take it up into the air, but then the bird cannot use these wings for anything else. If the bird wants to pick up something it must use its beak, it must use its head as an extra limb. That's why the head of the bird is only a servant of the trunk. But the forelegs have become wings, and that is the third secret of the bird. It cannot use its forelimbs for running, for picking up anything, but they are wonderful for flying.

And after the three secrets of thin hollow bones, air-sacs and wings, we come to the last and most wonderful secret of the bird: its feathers. Bird's feathers are not at all the same as hair, or as the fur on animals. Every bird's feather is something very complicated and special, and also something very beautiful. If you take a wing-feather you can see that in the middle there is the shaft or quill, and on either side of the shaft there is the vane. But the vane is made up of single barbs and each barb has tiny hooks that hold onto the hooks of the barbs next to it. It is because of these tiny, tiny hooks that the barbs seem to form one smooth surface, but the hooks can be separated. When this

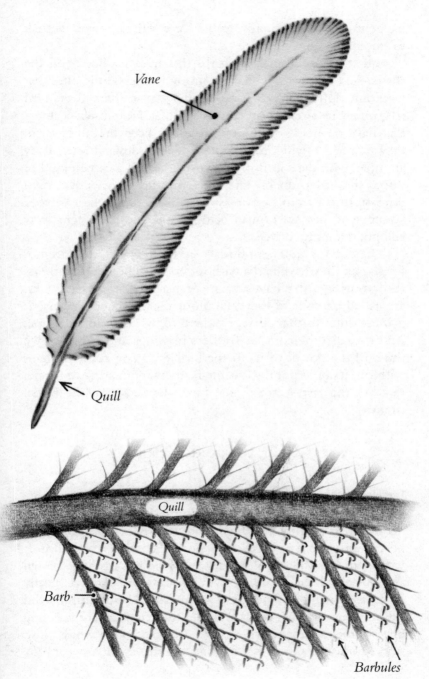

Vane

Quill

Quill

Barb

Barbules

happens on a bird, when its feathers get ruffled, it uses its beak to smooth its vanes again.

But there are also fluffy barbs that have no hooks on the shafts. And there are, under the strong feathers, little ones that have only fluffy barbs, they are called down-feathers. Every bird has, under its strong feathers, these fluffy little down-feathers that have no hooks. The down-feathers keep the bird warm: they are not for flying. You all know of eider-down duvets. They are filled with the soft fluffy down-feathers of a certain kind of duck, the eider duck. The duck mother plucks her own down-feathers from her breast to make a warm nest for her young, and then we human beings come and take them away and put them into duvets.

The down-feathers of birds keep them warm, they are not for flying, they have barbs without books. The other feathers, the strong ones that have a vane because the barbs have hooks, are called contour feathers. Contour means shape, because it is these outer feathers that give the bird his streamlined shape. And it is also the contour feathers in wings and tail that give the bird the power to fly. If the bird were covered with hair, with fur, it could not fly. It is the feathers, with shaft and vane, that are the last secret of the bird, the secret of the birds' flight.

Kiah, the golden eagle

We shall look at the life story of a golden eagle, a golden eagle that was hatched from an egg in Sutherland, in the north of Scotland. Its parents, the father and mother eagles, had built a nest — or eyrie as it is called — on the ledge of a high cliff. It was a sturdy nest, a strong nest that could not easily be blown away by winds or even gales. An eagle's eyrie is not as neatly made as a small bird's nest, it is not meant to be beautifully and skilfully put together, but it must be strong to protect the young ones on the mountains where sharp ice-cold winter winds blow. And so an eagle's eyrie is rather like a fortress, not a sweet little nest.

First the parent eagles put strong sticks as a foundation on the cliff ledge. Then they brought big branches from pine trees and it was the mother eagle who laid these evergreen branches so that they crossed and recrossed, like basketwork. And then the inside of this big and strong basket was lined thickly with heather. It was in such a nest that her young eagle was hatched from the egg; we shall call him Kiah, like the sound eagles sometimes make. When Kiah came from the egg he was not at all golden-brown as his parents were, he was covered with a grey-white down, and he was only a little bigger than a little duckling. And he had company; a little sister had come from the other egg in the nest. Eagles almost always lay only two eggs. And the first thing that our little eaglet Kiah and his sister did was that they started cheeping loudly for food. And the mother eagle was ready for them. The father eagle had killed a hare and the mother cut it up into little bits with her sharp beak and talons (claws), and she put the little bits into her children's open beaks.

But, as you will see from the whole story, eagles don't spoil their children. After a few days when the two little ones were stronger, the mother did not put bits into their beaks; she put the food down and the young ones had to pick it up themselves. But they were well looked after, both father and mother hunted all the time to have plenty of food, for young eagles eat a tremendous lot and grow fast. And when the winds became cold and sharp, the mother would sit on them and protect them and keep them warm under her wings. As the eaglets grew, they lost their fine down-feathers, and the real feathers began to grow. At first they looked like a hedgehog's quills, but soon the little spikes unfolded and became proper feathers.

Our little eagle, Kiah, would occasionally try to flap his wings, but the feathers were still soft, they bent and buckled, they were not yet ready to take the strain of the air pressing on them. In these first weeks all a young eagle's strength goes into growing feathers, and he is always hungry. And now that the parents just dropped the food on the nest, Kiah and his sister often fought each other and quarrelled for any bit that came down for them.

But as the feathers hardened, they became also lighter, and one day when Kiah flapped his wings he was suddenly lifted above the nest and hovered for a few moments — and then he came down with a plonk. After this experience Kiah was much more careful about flapping his wings, he still had to learn a lot about how to use his wings. After each meal Kiah would first scrape his beak against a branch to make sure that no bits of food were left on it and then he would take each long wing feather and run it through his beak to make sure that all the barbs were hooked together to make the vane whole. He would do the same for his tail-feathers, for the first thing an eagle must do is to look after his feathers; his whole life depends on them.

And now Kiah began to try flying. At first it was just a flapping of wings that carried him a few feet above the nest and let him hover a little — and then he came down again. But he soon discovered that it was easier when a wind was blowing, he could raise himself against the wind and the wind would carry him up. But Kiah was not yet quite ready for something that happened when he was playing with the wind. He had only given a little flap, but it was a strong breeze that was blowing, it carried him higher up than he had wanted to go; ten feet, twenty feet, and Kiah was frightened. In his fear he flapped his wings again, but this only carried him higher up. And if the wind had been a bit stronger it would have carried him away from the nest, over the cliff edge and — as Kiah was not yet strong enough — he would have fallen down hundreds of feet and would have been killed. But luckily for him the wind slowly died down and he came down in the nest, not too hard, but terrified, his heart beating madly with fear. For the next few days when he flapped his wings he gripped the nest at the same time with his talons so the wind could not carry him away.

Now that he and his sister had begun to use their wings, they also became more interested in the world outside the nest. And so they began to use the wonderful powers of sight that eagles have. They could see their parents flying towards the nest when they were still several miles away and then they would make their "kiah-kiah" sound which was their greeting. When Kiah and his sister looked at a bird half a mile away they could see every feather on its body. No other creature in the world has such wonderful eyesight as an eagle.

Our own eyes have lids to close them, each eye has one lid. But an eagles' eyes have three lids, the first two lids are only closed when he sleeps, the third lid is transparent, like glass, he can see through it. And the eagle uses this third lid when he is in full flight, then the wind does not make his eyes water when he flies at great speed. In this way the wisdom of nature protects her children.

There came the time when their parents thought that Kiah and his sister must make the great step and really fly. And so for a whole day the parent eagles left the young ones without food. The next day the parents flew away and Kiah and his sister screamed for hours with hunger. At long last the father and mother eagle returned and the mother held a dead wood-pigeon in her claws. But she did not bring the pigeon to the nest; she flew over the nest in circles, keeping the pigeon in her claws. And the father eagle too flew above the nest and made little circles as if he wanted to say, "Come on, come on, fly, you little cowards!"

The two young ones stretched out their necks but the mother did not come near enough. She hovered just at the edge of the cliff and by then Kiah was so hungry that he hopped to the edge and tried to grasp the pigeon, and as he did so he toppled over and fell. Kiah was so terrified that he forgot all he had learned, he thrashed with his wings, which did him no good, he spun head over tail and he screamed with terror. In an instant father and mother eagle were beside him, they could not hold him, but their presence made Kiah gather his senses. Suddenly he spread his wings flat instead of beating the air and immediately the breeze that was blowing carried him. He had fallen halfway down the high cliff but now at least he was hovering, he felt himself carried by the air, just as we feel buoyed up in water.

But he had not yet learned how to steer his flight by using his tail. It is the tail — that can be spread out in a half circle — which makes the eagle go this way or that. And now the breeze was carrying Kiah towards the top of some trees and again he screamed with fear. But again the parents flew beside him and their cries told him what to do. But Kiah was in such a panic that he did not listen. And when the treetops seemed to rush towards him, Kiah beat with his wings against the direction in which he was going, and this just made him fall and drop right among the thick fir branches. He suffered no harm but he was very scared and he sat for a long time perched on a branch, not daring to let go.

From there he watched his parents trying to get his sister to take flight. The sister too, at first, just fell off the cliff and only saved herself — at the cries of the parents — by spreading her wings at the last minute, but she was only a few feet from the ground and so came down soon after. There she stayed at the foot of the cliff, and no matter how her parents coaxed her, she would not try to fly up again.

So on that day the parents brought some food to each of their children where they were. But the next day the flying lessons had to go on. Kiah was again tempted by some food held out by his mother to fly from one pine tree to the next, and to the next one a little farther away, and the next one still farther. And through these short flights he learned to use his wings and his tail feathers better and better, he became confident and lost his fear. His sister too was first made to fly up to a lower branch of the tree, then higher up, then from tree to tree. And so they both learned the hard way the great art of flying.

Even eagles made by nature for flight must go through hardships until they can fly. It is also the same for us human beings. Learning cannot always be a pleasure, but what we have learned will be a pleasure to do once we have mastered it, as the eagle enjoys flying once he has mastered it. And it's no use being afraid or not trying hard, for then we will never learn anything.

In time Kiah learned the great art of flying as only an eagle can fly. He would launch himself from the nest, really throw himself from the cliff, and make the first flaps with his wings. When we see this start of an eagle's flight it may seem clumsy and slow compared with the way a little bird gets into the air. But this is only the start of the eagle's flight.

But Kiah would not flap his wings very long, for he would soon meet a current of rising air. You see, the air around us is never really quite still. There are currents in the air as there are currents in water, and the currents in the air go up, or go down. Kiah could feel even the slightest current, and as soon as he found a rising current (near a cliff or mountainside there is often a upgoing current, an up-draught, as it is called) Kiah would spread his wings and let the rising air carry him upwards. To stay within the rising air, Kiah would fly in circles, all the while rising in a spiral. And so Kiah would be lifted higher and higher, just keeping his wings spread out and following the rising air.

At first it sometimes happened that Kiah made a mistake and suddenly came out of the current of rising air; then he had to use his wings and flap to get back into it. But he soon became so practised that he could stay floating in the current and was lifted up without moving his spread-out wings even once. This is the famous soaring of the eagle; he rises in ever-higher circles without moving a wing.

In this way Kiah could reach heights of thousands of feet, heights where human eyes could only see him as a tiny dot, if they could see him at all. And from this great height Kiah would sometimes see, with his wonderful eyesight, a mountain hare nibbling at a bush. And then would come the other amazing feat of the eagle's flight: the diving. He would close his wings and let himself fall headfirst, but by using his tail, falling in the direction of the hare. He would fall, he would dive faster and faster — the speed with which eagles come down in a dive is faster than the fastest car — and open his wings at the last moment, just before he touched ground. And his coming had been so fast that the hare had no time to notice it until it was too late and the hare was caught.

Kiah also learned to use the soaring and then diving when travelling. When Kiah wanted to go from his nest on the cliff to a mountain many miles away he would not fly towards it. That's what other, ordinary birds do, but not a golden eagle. Kiah would soar upwards on the rising currents of air and when he was high up he would aim himself at the distant mountain and come down there in a long glide. For the whole journey he had hardly moved a wing.

Sometimes Kiah and his sister and their parents would play games with the soaring and gliding and diving. One of the old eagles would fly up with a mouse or squirrel in his claws — his talons — and all the others would follow him. And when they were all high up, the old eagle would let the dead animal drop. And as it fell Kiah came diving after it and he could dive with such speed that he always caught it well in mid-air and long before it touched the ground.

And there is still one more trick Kiah learned, a trick to be used at the end of a dive. When Kiah came from some other mountain crag and was diving towards the cliff where his own nest was, he of course dived headfirst, which is like falling headfirst. But one cannot land headfirst on the cliff: that might break the neck. So what would Kiah do? In the last moment — so fast that the human eye can hardly follow it — he could make an incredibly fast corkscrew movement that brought him into a feet-first position. He would land on his feet, with just a little slither. That is how eagles fly, they are truly the kings of the air.

When Kiah and his sister were five months old, they were already as large as their parents. Kiah had a wing-spread of 2 m (7 ft) from tip to tip. His feathers were a dark brown with a purple shine or glow, with touches of white on the wings and their undersides. The head and neck were just beginning to get the golden colour from which the golden eagle gets his name. By now, at the age of five months, Kiah could fly as well as his parents, and he could hunt and fight. As we saw, eagle parents don't spoil their children, and Kiah's parents decided that the time had come when Kiah and his sister must fend for themselves.

And so one day the whole family, father and mother, Kiah and his sister, were flying home to their nest on the cliff. They were not diving this time but flying leisurely, circling above the nest before coming down. The parents flew down first, but when Kiah and his sister followed them, both old eagles suddenly turned against them, struck at them with their beaks, beat at them with their wings, and would not let them land or come into the nest. At first, Kiah and his sister did not understand. This was their nest and home, these were the parents who had fed them, protected them, helped them to learn to fly, why did their parents now drive them off? Sadly the two young eagles flew away and spent the night in treetops.

The next morning Kiah saw his father fly out on the daily hunt and he flew up to join him. But as soon as the father eagle saw him, he dived down on him and Kiah only saved himself from being ripped by his father's claws by quickly swerving. But the father came screaming at him again and Kiah was so scared that he fled. And the father eagle came after him and drove him on and on until Kiah was miles and miles from the place that had been his home. At long last the father eagle flew back and Kiah came down on a tree and realized that his childhood was over. From now on, he had to be on his own, he had to hunt for himself.

But this was not quite so easy as Kiah thought. You see, every grown-up eagle has a certain territory, that means a wide circle of land where he hunts. And he will not allow any other eagle to hunt in his territory for there is not enough food for more than one family in one territory. And so several times when Kiah flew up to look for a hare or squirrel to hunt, other, older eagles came screaming at him and drove him away. But at long last he found a cliff that offered protection from the winds and this cliff became Kiah's centre from which he hunted. For more than two years Kiah made his flights from the cliff, and returned to it after a hunt, roosting in the branches of an old pine tree. As long as they are alone eagles don't have nests, they only roost as it is called, they rest in trees.

When Kiah was about three years old — it was springtime — he saw another eagle coming into his territory. Straightaway he flew up to drive the stranger away, but then he saw that this was a she-eagle and Kiah wanted to make her welcome, he wanted her to stay with him and to live with him. And so he began the beautiful courtship flight of the eagles. He flew around her in circles, he flew above her and suddenly dived down below, and she answered by flying beautiful loops up and down. And then they flew together, their wing tips almost touching. So Kiah had found his mate and they would stay together all their lives.

Together they built a strong fortress eyrie on the cliff and when the nest was ready the she-eagle laid two eggs and brooded on them. And while she was brooding, Kiah went hunting and brought the food. And sometimes the she-eagle would go for a flight, and Kiah would sit on the two precious eggs and keep them warm. And after one month there were two little eaglets, covered with white down.

And the whole story starts anew, where we began.

The Limbs

We shall look at the limbs of various animals, not just one. And we shall begin with the limbs of the animal we have just heard about, the bird. Of course, the most important limbs of the birds are the wings. The wings are really the forelegs or fore-paws. While other animals, like dogs and cats and horses, use the forelegs in the same way as the hindlegs — to carry the weight of the body, to stand, to walk — the birds' forelegs are wings that lift its body in the air.

And the feathers on the wings do not just grow higgledy-piggledy, a long one here a short one there. The feathers on bird's wings are arranged in a special way: the longest feathers — they are the feathers which are most important for flying — are on the outside, then the shorter feathers more on the inside, and there are little feathers which are only there to cover the shafts of the long feathers. It is this arrangement of feathers that makes the wings such wonderful instruments or tools for flying.

Now let us look at the bird's legs. They are not as beautiful as its wings are, compared with the legs of a stag or horse they look shrivelled. But they are just right for a bird. On a bird's legs

there are only four toes, three pointing forwards and one, like a thumb, pointing backwards. And they are rather long toes compared with ours. The foot of the bird is used quite differently from our feet. On the ground most birds do not walk, they must hop, even the eagle can only hop when on the ground. Their toes are not for walking on; they are mainly for perching on the branch of a tree. When a bird is perching, then the long toes go round the branch and hold it with great strength. Even a strong wind cannot blow a bird from a tree, and even when a bird sleeps, his toes still hold the branch strongly. So the foot of a bird, with its long toes, is just right for the bird, it is the right tool for the job it has to do.

But precisely because the wings are just right for flying and the feet are just right for perching, they are not much use for anything else. When a young eagle falls from the nest, the parents can't save it with their wings or talons. When any little bird falls to the ground from a nest in the tree, the parent birds cannot carry it back up to the nest. Their wings and feet are each made for one thing and not for any other.

Now let us look at an animal that is the very opposite of the bird which lives above the earth, in the air and in the light. The mole lives under the earth, in the darkness. And the mole, which makes its way by digging through the earth, has very interesting limbs. The forepaws are again very different from the paws of cats or dogs. The forepaws of a mole look, at first sight, like clumsy little human hands, but the fingers are short and the fingernails have become very hard and long and pointed. In this way the hand or paw has become a wonderful tool for shovelling earth, it is a shovel that the mole is born with. And with these two shovels the mole can dig so fast that if you find it above ground it can dig down and be underground in a few seconds. For a creature that lives under the earth, the shovel-like paw is a wonderful tool. But just because of this the paw is very little use for anything else, especially because this paw is always turned outwards, so the mole could never carry something in its paws as the eagle can carry a hare. The mole's paw is a tool for just one job.

And now let us look at something we have seen often, the feet of a cat, or it could even be a lion because the lion is just a large kind of cat. The feet of the cat are, first of all, for walking, for carrying the body fast over the earth. But the cat's feet can do more than this: a cat can climb a tree — that's something dogs cannot do. Why is it that a cat can climb a tree? Because it has sharp claws, claws so sharp that they go into the bark of the tree and so give the cat a hold. If the cat would walk all the time with its sharp claws out, rubbing against stones, they would not remain sharp and pointed for long, but would become blunt and dulled. That's why cats can retract their claws, they can draw them in. So nature has again given the cat a wonderful tool, a wonderful instrument for climbing trees or for catching their prey; it has given the cat claws that can be sheathed when they are not needed. But these paws with their sharp nails or claws are again tools for one or two things only: climbing and catching. They cannot be used for swimming or for making a nest or for digging like a mole. The cat's claws can do some jobs and no more.

Now let us remember the seal's flippers: they are just right for swimming, for life in the water, but the seal cannot run like a cat, climb a tree like a cat, or dig like a mole. The seal's flippers are wonderful tools for one job and one only, but this one job, swimming, they do very well indeed.

If we now think of our hands and our arms, they are not like wings. We cannot fly with them; our hands are not shovels, we could not quickly dig a hole for ourselves with our hands; we cannot rip our food with our nails as the cat does; we cannot perch as a bird does. Our hands are not special tools for any particular job or task, but our hands can make any special tool we want. Our hands can make aircraft that fly faster and higher than birds. Our hands can make shovels that dig in the earth, our hands can make ladders to climb up as high as we want. Our hands are not a tool for just one thing, they can make tools for any job we want, and then our hands use the tools. We can hammer a nail into the wall, we can sew a button on, we knit, we can cut our food, write, draw and paint, we can build houses and bridges. It is just because our hand is not made for just one thing that it is free to do all things.

We work with our hands and the work we do is not just for ourselves but also for others. Think how wonderful it is that we use our hands when we shake hands in the morning, we show by shaking hands that we are friends, that we like each other. And when we pray we fold our hands, we use our hands even when we speak to God. That is why the human hand is the most perfect of all limbs, and we are the only creature whose limbs are more important than either the head or the trunk.

Part II

The Elephant

We have heard about animals that live in Scotland: seals, deer, eagles, snails, harvest mice, hedgehogs. Now I want to tell you about an animal which lives in warm countries, in India and in Africa. It is a very big animal, the largest of all land animals: the elephant.

That's the first thing that strikes you when you look at an elephant — the size of the creature. Elephants can grow up to 3 m (10 ft) high and have an enormous weight, up to 5 tons; that's about as much as sixty or seventy adults weigh. This giant creature is just about as long as it is tall, so seen from the side it fits into a square. But this large size of the elephant can tell you something very interesting. Long ago, there were many many large animals in the world, giant animals, and the elephant was just one giant among others. But the others have died out, and only the elephant remains. At the time of the last Ice Age, the earth had more varied and exotic animals than had ever existed before or would be seen again. This lost world of giants included a North American beaver as large as a black bear, an armadillo from South America almost 2 m (7 ft) long and an Australlian kangaroo 3 m (10 ft) tall. And the lion, which ranged throughout Europe, was much larger than its modern relative. And so the large size of the elephant reminds us of very ancient times on earth, of times when giant creatures roamed the earth, and among these giant creatures there were the ancestors of the present elephants.

And these ancestors of the elephant had something which the elephant has not got today. They had long thick fur and were even larger than today's elephants. This furry ancestor of the

elephant is called a mammoth. How do we know what the
mammoth looked like? At that time there were people who
lived in caves, and they painted the mammoth, the furry elephant, on the walls of their caves. In our time some of these
mammoths have been found frozen in the Siberean ice which
preserved them just as they were. So the size of the elephant
takes us back to very ancient times on earth.

As we saw, the elephant can be drawn in a square, and that is
because it has hardly any neck. The head sits right on the body,
the big head is just a continuation of the body. And the mighty
legs are like pillars, they have to carry the great weight of the
body. The body is the master, and the head and legs are its servants. And the trunk-body really is the master, for this enormous
body also needs an enormous amount of food and water. Every
day an elephant eats about 150 kg (300 lb) of leaves, branches
and grass. And to get this vast amount of food, the elephant has
to look for food and eat almost all day long. So there are some
disadvantages to being such a large size. You need a lot of food
and you have to spend all your time feeding that big body.
Elephants in a zoo don't eat quite as much, but they do eat a lot.
The big size also has advantages, for no animal dares to attack an
elephant. In Africa there are lions and leopards, in India there
are tigers, but a lion, leopard or tiger very rarely attacks an elephant; usually when they see an elephant coming, they get out
of the way. So the elephant, because of its size and strength,
hardly has to fear any other animal.

In fact, other animals have reason to be grateful to the elephants in the jungle. In these dense forests where the elephants
live, trees are close together and bamboo and grass grow up to
great heights so that one can hardly get through. But a herd of
elephants just breaks and crushes anything in its way and tramples it down with their mighty legs, making a path in the jungle.
Every day the elephants use the same path down to the river to
drink water, and the path becomes as hard as stone. So the elephants are the road-builders of the jungle, making paths which
other animals and even human beings can use.

Now, you might think that an animal as large and heavy as an
elephant could not be fast. The elephant cannot gallop like a

horse, but a charging elephant can move faster than an Olympic sprinter (40 km/h, 25 mph) and can keep it up for quite a long time. Another surprising thing about the elephant's walking or running is its sound. We might expect such a heavy creature to walk with a heavy, thunderous tread, but its foot has a soft padding, and so the elephant's walk is really almost soundless. Of course if the elephant tramples something down you hear the breaking of the tree or bush, but otherwise the elephant's walk makes less noise than we make.

★

The first thing that strikes us about elephants is their great size, and the next thing is the long trunk. The trunk is really the nose of the elephant, but that is not quite correct. The nose and the upper lip have grown together, and so the trunk is also the upper part of the mouth. Of course for the elephants the trunk is much more than a nose. If we were to watch a herd of elephants feeding in the forest, we would see them curling the tip of their trunk round a leafy branch and breaking it off. Then the trunk, still holding the branch, turns inwards and downwards and pulls the branch into the mouth. They use the trunk as we use our hands. The elephant has only this one arm and hand, the trunk, but it can make wonderful use of it. It is a very strong kind of arm which can break off branches that we could not even bend; it is so strong that it can tear out young trees by their roots. Sometimes, if two elephants have an argument, they hit each other with their trunks. This does not hurt the other elephant very much, but if a human being is hit by an elephant's trunk, they would be knocked over.

But we are still watching the elephant herd, and we see two elephants coming together who like each other, and what do they do? They put their trunks together intertwining them — for them it is like shaking hands or kissing. So they use their trunks not only to break things or to pick things up or to fight, but also to show they like each other. And over there, in the elephant herd, you see an elephant mother with a little one, and the little elephant is holding onto its mother's tail by its little trunk. A human mother takes her little child by the hand, but a little elephant takes its mother by her tail with its trunk. While the trunk is a very strong kind of arm and hand, the tip of this trunk is so delicate that the elephant in the zoo can take a sweet from our hand, or it can pick up a pencil from the ground. So this mighty strong trunk can do quite delicate work.

But the elephant herd we are watching does not stay among the trees, they move along a well-trodden path to the river. For elephants not only go down to the river to drink, they go there to bathe. Elephants do more bathing than we human beings do. They go bathing twice a day if they can, and if they can't immerse themselves in water twice a day they suffer, becoming unhappy and ill. And as soon as the herd comes down to the river you can see another use for the trunk, they suck water up in it and give themselves a shower with their own trunks, and they shower each other for fun. And of course the mothers shower their young ones with their trunks. That's something we can't do with our hands.

Yet the trunk is still a nose to smell things, and it is a far better nose than we have. While this herd of elephants is splashing in the river, rolling in the water, showering each other, swimming — they can all swim — there is always one elephant who stays apart and keeps its trunk in the air. Its fine sense of smell would tell it of any danger, like human hunters for instance. Sometimes in the dry hot season a river runs dry and the elephants must look for water somewhere else, but they can smell water when they are still 7–8 km (4–5 miles) away, and then they make straight for the new lake or river. Sometimes in the hot, dry season a fire starts in the grassland, and the dry grass burns very fast, but the elephants can smell it when the fire is still far away, long before anything can be seen, and so they move to safety. So the trunk is not only an arm or a hand but also a real nose that has a much finer sense of smell than we have.

But the trunk is still another thing. There may be one elephant who gets angry, and what does it do? It lifts its trunk and makes a terrible trumpeting noise through it. Or if a little one has lost its mother, it too trumpets through its trunk; it is a whining, wailing sound, and the mother answers, "I'm coming, I'm coming," also by using her trunk as a trumpet. You can hear from a human voice — even in a foreign language which you don't understand — whether a person is angry or sad or happy or excited. The trumpeting of the elephants which they do with their trunks is just like the human voice, it shows how they feel. Isn't it wonderful what the elephants can do with their trunks!

★

Two things strike us if we see an elephant, firstly the immense size, and secondly the trunk, which is at the same time a nose and an arm, and, you could say, the instrument of his voice. Now we come to a third thing: the tusks. With the African elephant both bulls and cows (as they are called) have tusks, but with Indian elephants only the bulls have tusks. The tusks don't look like our teeth, but they are teeth. Human beings have 32 teeth altogether, 16 above and 16 below. The elephant has far fewer; it has the two tusks and then two teeth above and two below, that's all. And only the four inside teeth can be used for chewing, the tusks are no use for chewing.

Elephants use their tusks for another thing. They eat not only leaves but also roots, and these they dig up with their tusks. The tusks are also a weapon. Bull elephants sometimes fight each other; it happens when two bull elephants like the same cow elephant. And in such a fight the two bulls not only hit each other with their trunks but also thrust at each other with their tusks, sometimes causing deep wounds where blood flows. But they never kill each other. The weaker one gives in and trots away.

The African-elephant cows also have tusks and they can be quite dangerous. If an elephant mother smells the least danger to her calf she will attack whatever it is, whether a person or a lion. The person or lion may have no intention of doing anything against the little one, but the mother elephant doesn't ask questions, she smells that there is someone about who should not be there and she attacks, goring the intruder with her tusks and trampling them with her feet. So it is not safe to go near an elephant herd when there are very young ones with their mothers.

The elephant has only four other teeth besides the big tusks — two above and two below — to chew things. In this way elephants are not as well off as we are with our 32 teeth. But in another way they are better off, for they change their teeth, not just once as we do, but every six years. It is really necessary, for chewing branches and roots wears the teeth out and they must do a lot of chewing, eating all that food. So the teeth are slowly worn down, but the wisdom that God has put into nature looks after the elephant, and as the old set of teeth is being worn down a new set is slowly growing. An elephant can reach an age of eighty years, and in that time it changes its four teeth about twelve times, while we have only one change of teeth.

But it is only the grinding teeth inside the mouth, which are changed. The big tusks outside are not changed, they keep on growing from year to year and the tusk of a very old elephant can weigh as much as 180 kg (400 lb), as much as three people together. These tusks are the reason why thousands and thousands of elephants were shot and killed, for these tusks are ivory. The ivory was carved into ornaments, into chessmen and little figures (especially in China), and billiard balls were once made of ivory. People paid a lot for ivory things and so hunters went out into the forests of Africa and India to kill elephants just for their tusks, for the ivory. And so over the years many thousands of elephants, especially in Africa, were killed. Nowadays it is forbidden to shoot elephants, but there are still poachers who kill elephants just to make some money. It would be a great pity if this wonderful creature were to be exterminated just because some people are greedy.

I mentioned one difference between the elephants in Africa and in India: the cows of the Indian elephants have no tusks. But there are other differences too. Africa is a less populous place and so the African elephant is taller, wilder, and their ears and their tusks are a good deal larger. India is a country that had cities, beautiful buildings and wise men long before we had them in Europe. India was a civilized land thousands of years ago. And the Indian elephant is a little smaller, has smaller ears and tusks and — the great difference — the Indian elephant can be tamed. It can become very friendly with human beings and even work for them. But the African elephant remains wild, rarely becoming friendly with human beings and will not work for them. So the elephants that carry children on their backs or do things in a circus are all Indian elephants, not African.

In India and other countries of the Far East, elephants are captured and tamed not for fun, but for work. The riders of elephants are called mahouts, and they train the elephants to carry heavy loads, especially timber cut in the forest, or stones for building, and Indian princes were once proud to ride on elephants. But before an elephant can be trained to do anything, he must first be captured. Catching wild elephants could not be done without the help of elephants that have already become servants of human beings. That is the strange thing: that tamed elephants become traitors to their own, they help their human masters against their own wild brothers and sisters.

Tembu, the Indian elephant

There was once an elephant who started his life as a wild creature in the forests of India. The elephant's name was Tembu and when he was born you would hardly have known he was an elephant, for his little trunk was quite short, it was just a rather long nose. And little Tembu was lying on the ground, not knowing how to get up on his legs. But his mother helped him with her trunk and he stood up, still a bit wobbly on his feet, but he could walk soon after he was born, quite unlike human babies.

For the first year of his life little Tembu always kept close to his mother and held on to her tail with his trunk when the herd moved about. He learned much from his mother, he learned that he had nothing to fear from bears or tigers but that there were small enemies that were a great nuisance, insects which settle in the many wrinkles of an elephant's skin and bite and cause an itch. Going into water helps, but only a little. But there were also little friends, birds who came and perched on the elephants and pecked with their beaks at the insects and ate many of them. So Tembu had to learn to be friendly to the birds and not to chase them away with his trunk. And he learned that when going into a river one has to look out for crocodiles, for a crocodile might try and have a bite at a young elephant.

There were many things Tembu learned. For instance, his mother warned him not to become a rogue elephant. Elephants are mostly quite peaceful animals, they don't often quarrel, but sometimes when they get old, one or the other of the elephants becomes very quarrelsome and starts fights with the others for no reason at all. When this happens, there finally comes a day when all the other elephants of the herd lose patience with the quarrel-maker and they all attack him, beating him with their trunks and thrusting at him with their tusks. They chase him away from the herd and he is never allowed to return; nor will any other herd allow him to join them. And so the troublemaker is cast out, wandering alone for the rest of his life. Such an elephant is called a rogue elephant. And the rogue elephant gets worse in time, attacking any living thing he sees, even killing human beings, until he is hunted and shot. It is not a good thing to become a rogue, that is what Tembu learned.

He grew up and was big and strong enough not to need his mother's protection any more, he was a strong young elephant in the herd. There came a time when there was heavy rain and the river where the herd used to drink rose higher and higher. At such times the elephant herd was in the habit of moving, of going further uphill, where there was not so much mud. But on their way uphill they were suddenly faced with a row of elephants, tame elephants with men sitting on their backs. The herd of wild elephants did not take much notice of the mahouts, they advanced, sure the other elephants would give way. But they did not, and they would not let the wild ones pass. Then the wild elephants tried to push them aside, but this did not work. So they fought, but the tame elephants had been trained by their mahouts, and they knew exactly where to hit, where to butt and thrust, and the wild elephants, sore and bruised, were driven back. But just as they turned to go back to the river, men came from the direction of the river, some carrying burning torches, others beating drums and kettles and shouting, making a terrible racket. Elephants can't stand fire and they can't stand a loud noise, so the wild elephants turned round again, but now there was a whole circle of tame elephants around them with keepers on their backs. So the wild ones were slowly driven towards a stockade, that is a big round fence formed by a strong wall of logs, so strong that no elephant could break it. Only the entrance gate of the stockade was open, and the wild elephants were pushed by the tame ones through the gate and into the stockade. The wild ones fought as best they could, but nearly every wild elephant had two tame ones against him, and so, in the end, bruised and tired, all the wild elephants, among them Tembu, were inside the strong fence and the gate was closed. So Tembu had lost his freedom through other elephants who helped their human masters to capture the wild ones.

Tembu had fought with all his strength, but the tame elephants had got the better of him, they had beat him with their trunks, they had butted him with their heads, and in the end Tembu was so sore and exhausted that he let himself be pushed into the stockade. He was so tired that he hardly noticed that in the stockade one of his feet was chained to a big tree.

The next day Tembu recovered a little and had a fit of rage. He trumpeted wildly, flapped his ears, tore at his chains, threw himself against the tree — but all that did him no good. And the tame elephant came and beat him with his trunk until Tembu calmed down. Then the chains were unlocked from the tree but left around his foot, and then, accompanied by two tame elephants, Tembu was taken to the river to bathe and drink, and then he was brought back and chained again to the tree. Tembu was hungry, but the first day he was still so angry that he trampled on the food given to him. But the next day his hunger was stronger than his fury and he began to eat.

A few days later he found that there was always the same tame elephant that came to him, a cow. And this tame elephant-cow did not beat him, she just parried his trunk every time Tembu tried to hit her, and when he calmed down she grunted in a friendly manner and so Tembu began to feel a little friendlier towards that tame elephant. Tembu also noticed that it was always the same man who came to him every day and gave him his food. When Tembu behaved calmly, without tearing at the chains and trumpeting, the man spoke in a friendly way to him, but when Tembu got angry, the man used a goad, an iron rod with a sharp point, which hurt even through Tembu's thick skin.

In time Tembu even came to feel a little friendlier towards the man who gave him his food, but he also learned to fear the painful goad. And then came the day when Tembu was allowed to go out with the tame cow beside him, and the mahout riding on the tame cow's neck. Whenever Tembu wanted to trot off on his own the cow just got hold of his trunk with hers and then Tembu obeyed and stayed beside her. And the mahout leaned over and scratched Tembu's head, which Tembu liked very much. So this first outing went very well. And after a few outings of this kind the man climbed across from the cow's neck to Tembu's neck and Tembu did not mind. Some time later Tembu learned to kneel down and so the mahout could climb onto his head. And so the friendship between Tembu and his mahout grew very slowly, but it grew.

After a year Tembu had learned to drag logs out of the forest to the river. The logs were tied together, quite a load of them, then the heap of logs was fastened by ropes on one side of Tembu's head to a strap round the load. Tembu pulled and pulled the load down to the river, while the mahout sat on his neck, directing him. And now the mahout hardly ever had to use his goad, nor was there any need for the cow to come along. Tembu had become so fond of his mahout that when he saw him coming in the morning he squealed with joy, held out his trunk to be stroked and almost danced with pleasure. And in the evening, when the work was done, Tembu wanted to come with the mahout to his little hut where he lived with his wife and children. Tembu knew he could not go into the hut, and he stood outside and snorted and grunted all night, but this kept the children awake. So Tembu was shown that he had to rest and sleep some distance away from the house. It was the mahout's wife who insisted on it, she wanted the children to sleep undisturbed. Tembu obeyed, but somehow he seemed to know that it was the mahout's wife who kept him away from the house and whenever he saw her he turned his broad back to her. But he respected his master; he had become a tame elephant.

The Horse

We saw that the elephant, in spite of its size, makes hardly any sound when walking, because the sole of the foot has soft padding. Now we come to an animal that is not nearly as large as an elephant but which makes quite a loud noise when it runs, because its foot is the very opposite of padded. Its foot is specially hardened, it is a hoof. And that animal is the horse. The horse's hoof is the opposite of the elephant's soft padded foot, and the clatter of horse hooves can indeed be heard. A horse is quite a different creature from the elephant. The elephant has hardly any neck, the horse's head is carried by a proud, beautiful neck. The elephant's body is short and squat, the horse's body is long. The horse has a long bushy tail, the elephant has a bare, funny little tail. The horse is really a beautiful animal compared with the heavy form of an elephant.

Elephants can move very fast, especially when they are angry, but normally elephants don't like to run, they prefer to move slowly, while horses enjoy a fast trot or even a gallop. And if you see a horse and an elephant side by side, the bigger animal seems more sleepy, the horse is much more awake. A horse hardly ever stands absolutely still: it moves its feet, its head, its tail; there is always some quivering movement in a horse.

The horse belongs to the great family of hoofed animals. We have heard of the deer, but the deer have two toes or hooves, and so have all horned animals like cows, goats, sheep or antelopes. The horse has only one hoof and so have its cousins, the donkey, the mule and the zebra. The horse really stands and runs on only one toe, the middle toe, which has grown very big.

The other toes are small and higher up, they don't carry any weight. The nail of the middle toe has grown right around it, and that is the hoof.

Sometimes, instead of a gate, a farmer will use a cattle grid to keep hoofed animals in a field. The bars of the cattle grid are far enough apart to let the legs of the animals go through and so they do not like to try and cross, but the bars are close enough to let farm vehicles drive across. (Though it has been known for a sheep to roll over and over on its back to get to the better grass on the other side of a grid!) The cattle grid would be no use at all for keeping elephants from crossing, because their big padded feet are quite different to hooves. The legs of a horse are rather slim for the long and heavy body when compared to the legs of elephants, but the slim legs of horses have great strength. They are so strong that horses even sleep standing on their legs, so you could say that their legs never rest.

The elephants can only move and even run in one way, first the two legs on one side, then two legs on the other side. But a horse is different; it can move in three different ways. It always starts with a hind-leg. When a horse just walks, it starts with the left hindleg, then comes the right foreleg, then comes the right hindleg, then the left foreleg. But, going faster, the horse can also trot. Now the left hindleg and the right foreleg move forward at the same time then the right hindleg and left foreleg move at the same time. And the third movement is galloping, which is very fast, and then the horse moves first the left hindleg, then — at the same time — the right hindleg and the left foreleg and then again the right foreleg.

And so, whether a horse moves slowly or quickly, the clatter of its hooves has a certain rhythm, a certain music. First it is 1–2–3–4 (like "Twinkle, twinkle, little star"), or it is 1–2, 1–2 (like "Bobby Shafto") or it is 1–2–3 (like "The Skye Boat Song" to "The Ash Grove"). In music it is called 4/4 or 2/4 or 3/4 and all dancing tunes are basically in one of these. Horses moved in these rhythms long before people started to make music. But you only have to hear those rhythms to know whether the horse is walking, trotting, or galloping.

And because there is already a kind of music in the horse's

walking or running, the horse is the only animal that can be trained to move in time to music, as they do in a circus, or with soldiers riding to the tune of a band. And there are some really special horses in Vienna, the Lipizzaners, which have been trained to make very complicated dancing steps in time to music. You can't get elephants to do that!

★

Horses have been helpers and friends of humans for a long time, many thousands of years, so long that many people could not imagine that there was once a time when horses were wild creatures. But there was such a time, long, long ago, when in the continent of Asia there were great plains covered with grass. Just imagine: you could travel for days, hundreds of miles, and there would not be any high hills or forests, very few trees, and as far as you could see there would only be grass. And in these grassy plains of Asia, herds of wild horses moved about. And the first people who came to these plains were hunters who hunted horses for meat, just as they hunted buffalo or deer. At first these people thought of horses as just another animal that could be killed and eaten.

Then some hunters saw foals drinking milk from their mothers, and as these hunters had no cattle they thought, "'It would be a good thing to have some mare's milk." So they made themselves lassos and caught some mares with their foals, then tied them to poles near their tents. (These hunters lived in tents as they were always moving about in the great plains). And they used the mares as we use cows, some milk was given to the foals but the rest of the milk the people used for themselves.

And the foals grew up, and when these nomad hunters wanted to move and struck their tents, they said among themselves, "We have always had to carry our tents on our backs or drag them along the ground as we move from one place to another. Why not try and put them on the horses' backs?"

The horses did not like it when the folded tents were put on their backs. They bucked and kicked. But the hunters tied the loads more firmly on the horses' backs and they held the horses down by ropes around their necks and, after a time, the horses gave in and carried the loads.

In this way the horses had already become very useful, and now the hunters no longer went and killed wild horses, it was much more useful to catch them and to make them carry loads. Then one hunter thought, "What would happen if I tried to sit on a horse and let it carry me?"

By now the hunters had learned that a horse has a very sensitive mouth, it feels the slightest pressure. So they put a bit, a bar of iron, between the upper and lower jaws in the horse's mouth, and on either side of the bit there were reins. Whichever way you pulled the reins the horse had to go, if it went the other way the bit would hurt too much.

The hunter who had had this bright idea called some of his friends who put a bit and reins on the horse and held its head down. Then he climbed onto the horse's back, his friends gave him the reins and stood well back. At first the horse stood still, it did not quite know what had happened, but there was a weight on its back, the horse did not like it, so it jumped into the air with all four legs and came down with a plonk. The man on the back was shaken, but he held on.

Then the horse bucked, it curved its back and jumped from hindlegs to forelegs, hindlegs, forelegs, hoping to throw the man off. But he was a strong man and he pressed his knees hard into the horse's side to hold on, and he did hold on. The horse now reared, it stood on its hind feet, but this too did not get rid of the man. By then the horse had worked itself into a lather, it was sweating, but it was still trying to throw the man off. It started to run, to gallop as fast as it could, but the man stayed on its back, and when the horse had run itself tired he pulled at the reins and made the horse go back to the camp. The horse was exhausted, it was wet with sweat and was shivering, but it knew now that it had found its master. It was tied to a pole, rubbed dry, given water and food and the rider stroked it. The next day the horse let him get onto its back without putting up any fight.

★

That's how the first horse was broken (as it is called), which means getting it used to carrying a rider. It is not natural for a horse to carry any burden or a person on its back, quite the contrary: the horse, like any other animal, wants to get rid of the burden on its back. And in the old days when the horse was broken by men of great strength it really was a breaking for it was the horse's own will that was broken so that it should only follow the will of the rider. It was a kind of battle between the will of the rider and the will of the horse. Only men with great strength could win such a battle that did not last long, a few days at the most. When they had won, when they had broken the horse's will, then anyone could ride that horse. Some cowboys in America still break horses in this way, but nowadays horses are usually broken in a different and far more gentle way.

In our time the breaking of a horse takes many months, and after the breaking there comes more training. Breaking only means getting the horse to the stage when it is willing to carry a rider. That is only the beginning, then the horse has to be trained for its particular task: racing, showjumping or drawing a carriage, these are all quite different things. But let us start with the first step, before anything else can be done, the breaking.

Usually horses are left to run about quite freely until they are three years old. A younger horse, a horse under three years, is not strong enough to carry a rider. So for the first three years of its life the horse lives like the wild horses of ancient times, running around and feeding in an open field with other horses. But it is also getting used to coming into the stable, to getting fed there and being groomed, that is, brushed clean.

Blackie, the horse

Now we come to the story of the three-year-old horse called Blackie. One day when Blackie came into the stable at the usual time, when he would find some nice oats in the trough, the groom had put the oats not in the trough but in a blanket which he held to Blackie's mouth. Blackie did not mind eating the oats from the blanket, and after a few days of this he was quite used to the blanket. The blanket just meant food for Blackie. And so Blackie was not upset when, at the next feeding time, the oats were in the trough again but the blanket was put on his back. And for a week or so, at every feeding time, the blanket was put on his back.

Then came the day when the groom put the bit in Blackie's mouth when the horse was eating. Blackie was not so keen to have a piece of steel in his mouth, but the groom talked to him quietly and stroked him and after a while Blackie got used to the bit. It also had to do with feeding, it seemed to him. He got used to having a bit in his mouth with reins hanging loosely around his neck.

Then came the day when the bit was put in Blackie's mouth, the reins around his neck and the groom led him very gently by the reins from the stable, just walking by Blackie's head and pulling the reins gently, and only for a short time. The next day the little walk with reins and bit was longer and Blackie found that the bit in his mouth could hurt if he tried to pull his head away from the groom or if the groom pulled and Blackie did not want to go. And in time Blackie got used to the bit in his mouth.

The next thing was that Blackie was led out into the field, with bit and reins, and then a long rein was tied to the bit, and this time not the groom but the trainer held the end of the rein. And Blackie could run or trot or walk, but only in a circle, for the trainer held the end of the long rein and would not let him go. The long rein is called the lunge rein and for many weeks Blackie was on the lunge rein learning to trot, canter or gallop as the trainer wanted. And then he had to do all these things with a blanket on his back.

Still no one was riding Blackie. Only after three weeks on the lunge rein something new came, a saddle was put on his back and tied with straps to Blackie's body. Again this took weeks, to get Blackie used to having a saddle on his back. One day the groom was out in the field and he brought Blackie some oats. Then, while Blackie was enjoying them, the trainer mounted the saddle. He only stayed a short while, Blackie was just standing and the groom held his head by the reins. Blackie was not keen on this, but on the first day the man soon came off. The next day he stayed longer and after more weeks the rider took the reins and Blackie walked with him on his back. As you see, it had all happened so slowly and gradually that Blackie did not even know that he had been broken, that he had slowly learned to carry a man and to obey every pull on the reins.

Ann in the prairie

We heard that many thousands of years ago horses were wild animals that lived in herds, grazing and running about in grasslands that went on and on. If you now let some horses run wild on grasslands, on steppes, they would very soon live like their ancestors did, in no time at all they would become wild herd-animals. There would be a large herd and one male horse, or stallion, would be their leader, usually the strongest and fastest of the stallions. That's how wild herds have always lived, led by the strongest and fastest.

A few hundred years ago in America some tame horses were let loose and became wild herds again. Somehow, no one quite knows how, a number of horses got away from the men who owned them; perhaps they belonged to Spanish soldiers killed in battle. These horses found grasslands that went on for hundreds and hundreds of miles. They had foals and the foals grew up and themselves had foals, and after a time in the grassland of America, the prairie, there were large herds of horses known as mustangs.

For a long time the horses on the prairie were quite undisturbed, for people did not care to live in or near the endless rolling grassy plains. But sometimes people travelled through the prairie. In those times there were no railways or cars, and if a family had to travel through the prairie they travelled in a big cart covered with canvas, a wagon, drawn by a pair of strong carthorses. They carried all their possessions with them and usually also had an extra horse, a packhorse that carried sacks of flour and other things they would need.

At one time a family — father, mother, two older boys and a little girl of five — were travelling in this way through the prairie. Their packhorse was an old grey mare called Susie, and the little girl, Ann, was very fond of Susie. When the family was on the move, Susie with the bags on her back, trotted behind the wagon and little Ann, sitting in the back of the wagon, used to give her sugar and stroked the head of the animal.

Little Ann wanted very much to be allowed to ride on Susie, but the father thought the child was too young and if the horse were to jump suddenly Ann would fall off and get hurt. But little Ann pleaded with her father and she said, "Look, Susie is just now carrying two bags of flour, one each side, they are so heavy that she just could not jump with them. But I could sit comfortably between the two bags."

And in the end the father agreed. Ann was put on the grey mare between the sacks of flour and so they moved on: the horses pulling the wagon, the father driving, the mother and two sons in the wagon, and behind Susie carrying Ann.

On such journeys one travelled every day from morning till evening and only stopped for the night. At first the mother and the boys often looked out at the back to see if little Ann was safe on the back of the mare, but they saw that there was nothing to worry about. The old mare sometimes stopped to nibble some grass and so got well behind the wagon, but then she got into a little trot and soon caught up with the wagon again. And little Ann shouted with joy when Susie moved a bit faster, so everything was alright.

It was about four hours later that the father stopped the wagon to give himself a rest and stretch his legs — and then it was discovered that the old grey mare with Ann on her back had gone, she just wasn't there. The father and the eldest son immediately unharnessed the draft horses, mounted them and rode back the way they had come, hoping to find the horse and the little girl. They rode back a long way, but did not see the grey mare or the child, all they could see was the endless sea of grass stretching before them, and they could not even guess which way the old horse had gone.

What had happened to Ann, riding on Susie, the grey mare? As was her habit, Susie had stopped to nibble at some particularly juicy grass and the wagon had got quite some distance away from her. But then, instead of setting out at a trot to catch up, the grey mare lifted her nose in the air and sniffed. Then she gave a loud neigh and trotted away from the track of the wagon out into the wild grasslands.

Ann cried out, "Susie, what are you doing? Go back, go back!" But the horse paid no attention. Susie cried for help, but the wagon was already too far away for anyone to hear her voice. There were no reigns for Ann to pull, and even if there had been she would not have been strong enough to make the horse obey her. For Ann, being only a little girl, it was too high on Susie's back to try and jump down, or even to climb down. All she could do was to cry and to hold onto Susie's mane as the horse trotted quite fast, further and further away from the wagon trail and out into the wild grasslands.

The little girl could not imagine where the horse wanted to go but Susie seemed to know very well where she was going and from time to time she made sure she was moving in the right direction by sniffing the air. For two or three hours they went on, the horse with the sacks of flour and the little girl on its back, and then little Ann could see where Susie was heading. At first it seemed like many dots in the distance, but as they came near she saw it was a herd of wild horses, hundreds of them. Not only could she see them but she could hear them, for their hoof beats were like rolling thunder. And then they were right among them, and Ann was more scared than she had ever been in her life when she was among these wild creatures, hundreds of wild horses, jumping, leaping, cavorting, rearing. Susie, the mare, seemed very happy to be in that crowd, she sniffed at them, they sniffed at her, and that seemed to be their greeting.

Little Ann could do nothing but sit up there, she had stopped crying since it was no use, there was no one who could help her. Yet, strangely enough, one creature did come, a white stallion, who was the most beautiful horse Ann had ever seen, a proud strong neck, legs that showed enormous strength and speed, a flowing mane and flowing white tail. And when the white stallion came, all the other horses made way for him, just as people would make way for a king. The white stallion was a big horse, so big that his head was higher than Ann, sitting on Susie's back.

Ann was scared stiff when the stallion put his muzzle to her, but he only sniffed, and then he took hold of her dress by the neck, between his teeth. He took her off Susie's back and he trotted a little distance to a stream, and there he put her very gently down. Ann drank some water, and she was so exhausted from all the excitement that she lay down there and fell asleep.

When she woke up it was already the dawn of the next day. She looked round in despair, what could she do? How could she find her family? She looked round, and saw Susie among the other horses. And now Susie came over to her and looked at her as if she wanted to say, "Come on, it's time to go back."

But Ann could not get up on the horse's back, she was too little, nor were there any stirrups. But again the white stallion came and he got his teeth into the neck of her dress and lifted her up. And then Susie gave a long loud neigh as if to say goodbye, and set off at a good trot.

By this time Ann's people had searched a whole day and most of the night for her and had given her up as lost. Imagine their joy when they heard — clop, clop, clop — the hoof beats of a single horse and saw Susie coming with Ann on her back. They did not punish the horse; they were too pleased that she had come back with Ann. But, as you can imagine, there were no more rides on horseback for Ann on this journey. Ann was only five years old when this happened to her, but she remembered it all her life and she specially remembered the beautiful white stallion who had helped her.

The Bear

The elephant and the horse, although they are so different, have one thing in common: they eat plants — grass and leaves. So do cattle, goats, sheep, antelopes, giraffes. They are all herbivores, which means plant-eaters. They don't kill other animals for their food; they feed on plants only. On the other hand there are carnivores, which means meat-eaters — animals like lions, tigers, wolves, foxes, otters who cannot live on plants and must kill other animals for their food. The teeth of plant-eaters are quite different from the teeth of meat-eating animals. Herbivores have teeth which mainly grind the leaves and the grass, teeth like our back teeth, they are blunt, broad teeth. But the carnivores have sharp pointed teeth, for with their teeth they must bite, kill, tear the meat into smaller bits. And it is the same with the limbs of animals. The herbivores, the plant-eaters, have hooves, or like the elephants, padded soles, but they don't have claws. But lions, tigers, wolves, need claws in order to hunt and kill their prey and also to help tear it up.

We human beings have teeth for both kinds of food. We can chew and grind plant food and we also have sharp teeth for meat. We are omnivores: we can eat all things, plants and animals. Now there is another creature in the world that can eat plants and animals, a creature that can eat fruit and meat, and that is the bear. Bears enjoy eating berries, they like honey, but they will also eat squirrels or rabbits or fish.

There is still another similarity between bears and human beings. Bears can and do stand up on their hind legs quite naturally, they can and do walk upright, and they use their forelegs for fighting, hitting, striking, as we use our arms. The forelimbs are really much shorter than the hind-limbs and so, when the bear trots on all fours, its hind parts are higher than its head, as it is with us when we try to walk on all fours. But the bear goes down on all fours when it wants to run, when it wants to move quickly. It can stand on its hind legs or walk slowly on them, but it cannot run on two legs, for that it has to go down on all fours.

So you see, even the bear — who can quite naturally stand and walk upright — has to be on all fours when it wants to run. Now when the bear stands upright it can use its forelimbs quite cleverly as a weapon, the claws are sharp and the bear has a terrific strength when it strikes. It can use its forelimbs for throwing fish out of the water; it strikes the fish and makes it fly out of the water onto the land. Or sometimes one can see little bears having a wrestling match and they wrestle just as two boys would wrestle. But the bear cannot use its forelimbs to make anything; they are not like hands. The bear's forelimbs, when it is upright, are a weapon to fight, to strike, to wrestle. And when it is on all fours the forelimbs are just feet to move fast. So even for the bear, the forelimbs are just servants of the trunk. They help the trunk to get food, to run away from danger, but they cannot make things. The bear, too, is a trunk animal.

The bear is the strongest and biggest of all the land carnivores. A big brown bear can weigh over a ton, but this big strong creature has something in common with another little animal which we have heard about: the hedgehog. Like the hedgehog the bear hibernates, sleeping through the winter. And this is one of those things which are rather mysterious. A human being could not be without food or water for more than a few days, even if they rest and sleep. But the bear is in its winter sleep for many months, and in these months of sleep it eats nothing and drinks nothing, it is quite strange. In any case, when the winter comes the bear acts very much like the hedgehog: it looks for a place where it can hide, like a cave or the hollow trunk of an old dead tree, and there it makes itself a soft mattress of moss, leaves and grass, lies down and goes to sleep for many months on end.

So you see bears are strange creatures, they eat plants and animals as people do, they can stand and walk upright like people they climb trees like cats, they can swim in water like dogs, and they hibernate like hedgehogs. Bears can be good-natured and placid like cows, but they can also become furious, and then they are terrible killers like lions or tigers. The bear is like a mixture of many animals.

★

In Britain we can see bears only in the zoo. To see them in the wild you would have to travel quite far. About nine hundred years ago there were still big brown bears in Scotland, but, as the forests were cut down to make room for fields, the bears were killed off and none of them are left. In Europe there are still a few wild brown bears in Russia, and there are more in Siberia and North America. These countries still have great forests.

The brown bear is a forest animal and it can only live where there are forests, but they can become quite dangerous to farmers who live near the forest.

There was once a Russian farmer who encountered a bear. The farmer was an old man and one day in autumn he was all by himself because his family had gone with a herd of cows, driving the herd up to a distant hill where there was still good grazing. So they were all gone and the old man was left by himself. He was sitting in the house when he heard the hens outside screeching and running and so he went out to see what was the matter. He could not see anything in the courtyard, but a little further away was the farm's orchard.

When the farmer came to the orchard he saw a bear-cub trying to climb a plum tree. The plums were just about getting sweet and ripe and the farmer had been looking forward to the plum jam that would be made from them, so he was very angry at seeing the little bear trying to get at the tree to help itself to the plums. There was a stick lying in the grass and the farmer took the stick and started to hit at the bear cub, which was just about halfway up the tree. The little bear squeaked and tried to come down the tree backwards.

Just then the farmer heard a grunting sound behind him; he turned round and saw a she-bear running towards him from the forest. No doubt she had heard the cub's squeak and was coming to fight for it. The farmer was terrified, he knew the bear would be upon him before he could reach the house and so he started to climb the nearest strong tree — of course not the plum tree on which the cub was. By the time the farmer had got to the top of the tree the she-bear was right underneath. And was she in a temper! She bit into the trunk and she tore furiously at the bark with her claws. Then she stood up on her hind legs and began pulling and breaking all the branches she could reach with her forepaws. The man was hanging on at the top of the tree, yelling for help, but there was no one to hear him. He expected that any moment the great beast would come climbing after him.

Fortunately for the man the bear suddenly changed her mind and turned to look for her cub, which had in the meantime climbed the plum tree and had started to eat the plums. It seems that she decided the youngster needed a lesson. She went to the plum tree, gave a grunt, and the cub came quickly down. Then the bear cuffed the little cub's ears until it ran squealing back into the forest, followed by his mother.

The farmer stayed up in his tree for a long time, he was afraid that the she-bear would come back and attack him. However, the bear did not return, and the farmer finally came down and rushed to the house. You can imagine how excited his family were when they came back in the evening and heard what had happened.

Gerard and Denys in the forest

The encounter of the farmer with a brown bear ended well for both bear and man. But it is not always so. Only a few hundred years ago wild brown bears were to be found in many parts of Europe: in Germany, for instance, only a few miles out of town. It was about three hundred years ago when two young men travelled on foot from one town in Germany to another. In those days travelling on the highway was never quite safe. There were not many travellers and you could walk or ride for hours without seeing another soul. But then, if you did see people, you might not want to see them at all because they could be a band of robbers who would take from you all you had. And, as a part of the road went through dense forest, there was always a chance of meeting a wolf or a bear. So it is not surprising that one of the two young men, Gerard was his name, carried a crossbow with him. A crossbow, which shoots steel shafts or bolts, not arrows, was a useful weapon in those days.

So these two young men, Gerard with his crossbow and his friend Denys, were walking on the road. They were now entering the part of the road that ran through the forest. On either side of the road trees and bushes stood so close together that you could hardly see into the forest that stretched behind. And Gerard said to his friend, "Now keep your eyes and ears open. People say there is a robber hiding in the forest and we don't want to be taken by surprise."

So the two friends walked on, looking right and left, sharply alert for any movement or noise which might betray an attack. For a while there was only the whisper of leaves in the wind and then, suddenly, something moved in a bush ahead of them. Gerard, his crossbow at the ready, shouted, "You there, come out and show yourself or I'll shoot!"

Something moved in the bush but did not come out, so Gerard aimed and shot his bolt at the bush. And the next moment a young bear cub leapt out and fell down on the road, the bolt sticking from its side. When the young men came near they saw the animal was already dead.

"What a pity," said Gerard, "but you know, Denys, the fur will be useful for something. Do you have a knife to cut the creature up?"

He had hardly finished speaking when they both heard a rushing sound coming over the dead leaves on the road. They turned round and saw, about sixty paces away, a big brown she-bear running at them. When they saw the big beast the two youngsters lost their heads, Gerard dropped his crossbow and ran for the nearest tree while Denys ran for a tree on the other side of the road. The she-bear first sniffed at the cub lying in the road. When she found it was dead she gave a terrible howl. Then she made for the tree in which Denys was still climbing. She tried to reach him with her claws but he was already out of reach. With fury the bear tore a few chunks out of the tree and then came climbing after him. Denys had a short dagger with him, but he knew that a dagger was useless against such a mighty creature as this bear, and he thought his last hour had come.

When Gerard saw the mortal danger his friend was in, he came down from his tree as fast as he could, picked up his crossbow, charged it with a bolt, ran to the tree and shot the bolt into the broad back of the bear. The beast took no notice, but when a second and third bolt struck into her she gave a furious yelp and, instead of climbing further up towards Denys, climbed down again to deal with her attacker. Gerard saw her coming down. He could not fight her with the crossbow and so he dropped the weapon and the bag of the bolts, ran for his tree and climbed up again. And there was the she-bear, mad with fury and climbing after him. Gerard had reached a big forking branch and climbed out on it as far as he could. He hoped the heavy beast would not follow him, but the bear seemed to be as surefooted as a cat and came out onto the branch.

By now Denys, seeing his friend in danger, had come down from the tree. Now he took the crossbow and aimed at the bear. Two more bolts struck the animal, this time in front. She stopped, and then fell down from the tree with a terrible crash. Even then, after that heavy fall and with mortal wounds in the body, the she-bear was not quite dead and Gerard had to kill her with a thrust of the dagger into the heart. If the two friends had not helped each other, the bear would certainly have killed them both.

Jessie and the grizzly bear

The grizzly bear is the largest and strongest of the brown bears. There was once a mountain pony, a small but strong horse that is used by hunters in the mountains of Canada. She was a mare called Jessie, and her master, Mark, was a hunter. Mark was out hunting in the mountains of Canada, riding the sturdy mountain pony, Jessie. For several days Mark had watched a herd of wild mountain goats, he wanted to shoot one of them to get some meat and to get a pair of their beautiful strong horns. Having watched the mountain goats from a distance for long enough he had decided from which side he would approach them to get as near as possible.

Of course one can't get very near to these wild creatures, but near enough for a rifle shot. So when Mark had ridden up the narrow mountain path, he saw a wild goat near a cliff. He took his rifle, aimed and shot. The goats scattered in all directions, but one of them fell, and it fell down the steep cliff. But the goat did not fall all the way. Halfway up the cliff there was a ledge and the goat fell onto the ledge and there it stayed, motionless, dead, but not much use to Mark. What could he do?

Then Mark remembered he had a long, strong rope with him. If he went up to the cliff, made the rope fast to a rock and let it hang down the cliff, it would be just long enough to reach the ledge with the dead goat. And so Mark dismounted, left his pony and his rifle some distance from the cliff, and went to the clifftop with his rope. As he had planned, he fastened the rope to a rock and let the rope dangle down to the ledge. And then he climbed down the rope. Once he got to the ledge he would tie the goat to the rope, climb back and pull the rope up. That is what he would have done. But he had just got to the ledge when he heard an angry "whoof" above, he looked up and saw an enormous grizzly bear, right on top of the cliff, and glaring down on him with every sign of fury. No doubt this was a she-bear. Her cubs must be near the cliff in some cave and the shooting and all the goings-on had upset her; she must think somebody was attacking her young.

Mark could do nothing, he could not climb back when the she-bear was waiting to crush him with a blow of her paws, and he could not get down from the ledge, as there was a straight drop of 15 m (50 ft). Nor could he stay on the ledge, for there was an icy wind blowing and he could not last long in the cold blast. Mark was giving himself up for lost when he heard his pony neigh. "Poor Jessie," he thought, "now the bear will go for her, I only hope she will be fast enough to escape."

The grizzly did go after Jessie in his loping gait, but Jessie did not run away. She did something quite unexpected; she turned her back on the bear, but stood quite still. The bear liked that, coming on a prey from behind was an easy kill. But when the bear was near enough to reach for Jessie's flanks with its paws, Jessie kicked — she kicked with both hind hoofs, hoofs that were iron-shod, and the kick landed on the bear's head. A human head would have been shattered, but bears have a harder skull and the bear only felt a hard blow. She stood up on her hind legs — then Jessie kicked again and the hoofs struck the bear's chest and threw her over. And that was enough for the bear, with a squeal she made off.

Mark could hear it all happen from his ledge. He quickly climbed up on the rope, he did not bother about the goat, he did not bother about the rope. He hurried to Jessie, who had saved him from a desperate situation, mounted, and rode away as fast as he could. He had no wish to wait for the bear to have another try.

★

We have seen the similarity between human beings and bears who have the ability to stand up and even walk on their hind legs. However, bears can only use their forelimbs for fighting and killing, they cannot use their forelimbs to make anything; they are not like hands. Only the human being has the gift of hands which can create things and give things, and so with the human being the hands are the servants of our head and heart, and not like the animal's limbs which are just servants of the trunk.

The Lion

Now we come to the animal which is called the king of the beasts — the lion. The lion is actually one of the big cats. It is a very strong animal and when it is hunting or fighting it can be terribly fierce, though at most other times it is restful. When people want to describe a person who is bold and fearless, they say that person is "like a lion." There was once a famous king who was such a courageous leader that people called him the Lionheart. The grown-up male lions have a mane (no other cats have a mane) and, together with their dignified, confident manner and their broad heads, this mane makes them look so magnificent that you can understand why people call the lion king of beasts.

The lion is a very muscular animal: it can run very fast and with one big leap it can cover a great distance — the length of a bus. One blow from its big paw would be enough to knock us flying and its jaws are so powerful that the back teeth can bite through a bone as easily as we could bite through a stick of celery. When a big lion gives one of its thundering roars, it is so loud that it can be heard several miles away, warning all others to beware. But even with this great muscular strength, the lion is never lumbering or clumsy; it is always graceful and supple in the way it moves. It can show terrific force and tension, but also complete relaxation.

The lion is actually a big cat, but while all other kinds of cats — tigers, leopards, wildcats — prefer to live alone most of the time, lions like to live in big families. A family of lions is called a pride, and there could easily be twenty animals in a pride: say two lions, six lionesses and all their cubs. There are always more lionesses than lions, but the big adult males with their flowing manes are always the leaders; they must play the main part in defending the pride and its territory — not only from dangerous intruders like hyenas, but also from other lions who sometimes try to take over as leader. When this happens there can be very ferocious fights, for the adult males are the biggest and strongest of all the pride.

During the heat of the day the lions like to lie down together in a group. You might hear them grunting, snarling and roaring as they move about and chat to one another. When we meet a friend or relative we shake hands or give a hug, but when lions greet each other they rub heads and necks, they nuzzle. A pet cat spends a lot of time cleaning itself — licking its fur with its rough tongue. Lions do the same, except that they also like to lick one another. After all, it is impossible to lick your own neck and head properly, so it is good if someone else can do it for you! The lions lick one another and the mothers lick their cubs. By doing this they are cleaning and grooming, but also passing around the particular smell of the pride, and the lions like this very much. At the same time the mothers will give their cubs a loving nudge with their muzzle, then you can hear deep purring and miaowing, and sometimes also a little hissing and snarling. The lions can also show something of their feelings in their faces; altogether they are very "expressive" animals.

If they can find any shade the lions prefer to lie there — perhaps under a tree or bush. Then they will stretch out or curl up — sometimes they will lay their head on the lion next to them and close their eyes to rest. The lionesses and their cubs spend a lot of time together without the male lions; the lionesses in the pride are all related to one another — sisters, daughters, aunts — and the older ones work well as a team. The whole pride does not stay together all the time. The lions separate into smaller groups and travel through their territory for a while, finding adventures and looking for food. Some lions even choose to go off by themselves or as a pair.

Nearly all wild lions live in Africa in big wide grasslands known as savannah. The savannah is mostly rather flat and goes on for mile after mile, there are just a few trees and bushes here and there but no woods or forests, not even a group of trees together. In the savannah there is a rainy season and a dry season, but much of the time it is the dry season and very little rain falls. Then it is as dry as a desert and the grass is not green and juicy as it is here, it is tough and brownish-golden in colour, just like the colour of the lion's coat. Here and there a few water-holes are left over from the last rainy season and many kinds of animals go to them for precious water during hot dry times. The lions also go to the waterholes, not just for the water, but to find something to eat. In the savannah live big herds of hoofed, grazing animals: antelopes, wildebeest, buffaloes, zebras, and only much smaller numbers of hunting animals or predators: lions, leopards, cheetahs. There are also lots of birds and other animals, like snakes, monkeys, hyenas, giraffes and rhinoceroses.

Hunting for food

The lions like to rest when it is hot — stretched out to their full length or curled round in the way that cats do. The little cubs do not rest so much, they love to play — chasing, catching and tussling. They often scamper around for a bit while the big lions are relaxing in the heat. When people see lions sleeping in the heat of the day they sometimes think, what lazy beasts these lions are; they just lie about for hours, relaxing, sleeping and taking it easy. It is quite true that they spend the biggest part of the day resting and sleeping, but when it is time to get some food for the family, when they go hunting, it is another matter altogether. When the sun goes down and the heat gets less, the time has come to get active.

The lions get up and have a good yawn, they thoroughly stretch their muscles and flex their claws: they are getting ready for action. They can actually hunt at any time, but they prefer the cooler times of day: dusk, night time and especially dawn. Lions and all other cats can see very well in dim light, even when it is so dark that we could hardly see anything at all. They also have very acute hearing and an excellent sense of smell. The best hunters are the lionesses — they are smaller, faster and more agile than the big males. The whole family, the whole pride, depends on their skill, patience and concentration when stalking their prey.

If you see a lioness going after an antelope or wildebeest it is hard to think of any other creature which can concentrate so completely on what it is doing. She focuses all her attention on the job in hand, not looking right or left, not being distracted by anything else that is going on. Her eyes are so fixed on her prey that for her it is the only thing she sees at that moment. Lions have padded feet, they walk almost silently, and the way the lioness creeps up quietly on the grazing herd is amazing to see. She crouches down low so that her belly brushes the ground and, with legs bent, she slowly and cautiously moves forward, first one leg, then another. But all this happens so smoothly that she just seems to glide in slow motion through the grass, and if the zebra happens to look in her direction she will freeze, not moving a muscle, until it looks away again. There are no sudden movements, there is no sound as the lioness just inches closer and closer to the zebra. She is very patient and can go on stalking in this way for hours.

To walk in this crouching position for any length of time needs great strength and agility, but, as we saw, the lion is a strong and powerful creature. So you might be wondering: if the lioness can run so fast, then why does she have to go to all this trouble? Why doesn't she just chase the antelope and catch it? Why does she spend so long stalking in this way? It is true that lions can run very fast and take great leaps, but they cannot keep it up for long. If you think of human sprinters running in a short-distance race, they also go very fast but after a very short time they are exhausted and gasping for breath. The lions are a bit like these muscular sprinters; although they are very good at giving a sudden burst of energy, they can't keep it up for long. But hoofed animals like the zebra or antelope have greater staying power. An antelope is not as fast as a lion but it can go on running much longer, it has greater endurance. Just think how long a horse can go on galloping; it goes on and on for miles. So hoofed animals are more like long-distance runners — not so fast in the sprint but better over long distances. And so it often happens that when a lion chases a zebra or a wildebeest, the hoofed animal escapes unless it is caught quickly. That is why it is important for the lion to get as close to its prey as possible by stalking before it sprints forward in a last sudden dash.

Often the lionesses work together; they hunt as a team, they fan out and form a big circle round a herd of zebras or antelopes. For these grazing animals the grass is food, it is something to eat, but for lions the very same grass is not food, it is cover — something to hide behind when hunting. Some of the lionesses will be hidden in the long grass and then those on the far side of the circle will chase an antelope towards their hidden companions. Then a hidden lioness leaps out and catches the prey while the others come running to help. They leap onto it, holding on with their front paws, and soon kill it. This will give a good meal for the whole family, for the whole pride. A lion can eat a huge amount at one meal, as much as 25 kg (50 lb) of meat, but afterwards it feels very sleepy and lies down for a long rest while it digests its food.

Sometimes it seems to us as if nature is cruel, but it only seems so. Often, when a lion catches a zebra, people have noticed that the zebra "gives in" — it stops struggling. It has also been noticed that, if the lion is driven off in time, a seemingly dead animal can run away unharmed. The Scottish explorer, David Livingstone, once had the experience of being attacked by a lion, but he was lucky. He was rescued just in time and lived to tell the tale. What he told is very interesting: he said that the growling lion took him in its jaws and shook him as a terrier dog shakes a rat, or as a cat shakes a mouse. Of course being grabbed by a lion was a terrible shock, but then a strange thing happened. In the moment that the lion shook him he felt no more fear and no more pain, it was as if he was dreaming and yet wide awake. He felt quite calm and did not think of struggling, yet he was aware of all that was happening. In the next moment his friends rushed up, drove the lion away and he was safe. But he always remembered his close encounter with the lion and wondered if, when a lion catches an animal, the animal too might feel no pain or fear.

It looks as if the lions and zebras, the hunters and the hunted, must be enemies, but it only looks like it. If there were no hunting animals then there would soon be too many grazing animals and then there would not be enough food for them all to eat. If this happened then many of them would die of hunger. But life is also difficult for the lions on the savannah, there are many times when they cannot catch anything and they go hungry, and so there are never too many lions either. So you see, the wisdom which is at work in nature keeps things in balance.

Lion cubs

It is the mother lions that take responsibility for bringing up the cubs. When it is time for a lioness to give birth to her cubs she will find a quiet, sheltered place away from the others. This place is called her den. It might be among some rocks or bushes or it might just be a comfortable hollow hidden among the long grass. When the cubs — there might be two, three or even four — are born, they are small and helpless, their eyes are not yet open and they have no teeth. It will be three weeks before they are able to walk. Grown-up lions do not have spots or stripes, but the little cubs do have a soft pattern of spots, which only gradually fades as they get older.

The mother lion makes a new den every so often because the savannah can be a dangerous place. For example, there are hyenas about, and they might think that a little lion cub would make a very tasty meal if they could quickly dash in and grab one. After all, the lioness cannot stay with her cubs all the time, — she still has to go hunting for food. When the lioness wants to move her cubs from one place to another she picks them up in her mouth by the scruff of the neck and carries them to a new den. Her jaws, which are so strong and fierce when she is hunting, are now very gentle and the cubs find it quite comfortable to be carried in this way, they just relax and do not struggle to be put down. When the cubs are big enough — about two months — the mother lions of the pride take them to a kind of nursery so that they can help each other in bringing them up. And the little ones will be just as happy in the care of one of their aunts as they are with their own mother. At first the cubs live only on milk and they will already be six weeks old before they are given a little meat to eat. By three months their teeth are coming through and they gradually get more meat, but they are six months old before they are weaned, before they no longer get any milk and live only on meat.

When they are old enough the mother will take the cubs to an animal which the pride has caught and killed so that they learn to get the meat off the bones for themselves. They nibble with their front teeth and use their rough tongues to lick the meat from the bones. When they are one year old there comes the special day when they are allowed to go on their first hunt with the lionesses. At first they just watch, but after a few times they gradually learn how to keep hidden and quiet, how to concentrate hard, and how to stalk the prey. But it will be another year before they have learnt well enough to be of any real help on a hunt. The cubs also learn a lot when they are just playing with one another; chasing, catching and biting, though they also learn not to hurt one another by biting too hard.

When the cubs are about three years old the "girl" lions stay with the pride but the "boy" lions start to grow a mane and the time comes when they will have to leave the pride. One day the parents start to growl and chase the young males away — now they will have to look after themselves. This may seem hard, but it has to be so, because one day they will have their own family and they will be leaders of their own pride in a new territory somewhere else on the savannah.

Nowadays many lions live in big wildlife parks in the savannah. These are places where the animals are protected and no one is allowed to cause any harm — the Serengeti in Tanzania and the Masai Mara in Kenya are two famous wildlife parks where you can see lions today. But why do we need special protected parks for the wild animals? At one time, long ago, humans played only a small part in the working of nature and just fitted in with the animals and plants. But now, because of our farming and fishing and all our industries, we human beings have a big effect on what happens in nature. Sometimes this is a good thing and we help the plants and animals, but all too often we upset the balance of nature and unintentionally cause a lot of harm. For example, there are not nearly as many lions today as there used to be. More and more it is up to us to protect animals in the wild so that they can go on living as they have done for thousands of years, and today there are many people who are working in national parks and also in many other ways so that this can happen.

Buddy, the Guide Dog

The lion, the king of beasts, is a relative of the pet cats which some people keep at home. Another animal which we like to keep in our homes is known as "man's best friend." It is the dog, and of course all dogs are relatives of another wild animal, the wolf. Wild cats (except for the lions) live alone and hunt alone, they like to be independent, but wolves live in packs — and so a dog makes a very different kind of pet to a cat. A dog will treat the people who live with it as its pack and it needs to know who is the leader of the pack. It will be a much happier animal if it knows that its owner is the leader, and is in charge and decides what is happening. So dogs take to training more easily than cats do, and they enjoy working with people in different ways as hunting dogs, retrievers, guard dogs, sniffer dogs, sheep dogs, and — one of the most wonderful of all — as guide dogs for the blind.

Human beings and dogs have lived together for thousands of years, but it was only about a hundred years ago that some people thought guide dogs for the blind would really be a possibility. When the idea was first suggested many people thought that it must be a joke — just imagine a poor blind person running along holding on to the lead of a dog, being pulled this way and that as the dog chased other animals or stopped to sniff at every tree! But there were a few people who believed that with the right dogs and proper training something could be achieved.

There was a blind man called Morris Frank who lived in America and who started the first guide dog training school there. Morris was not born blind; he lost his sight because of accidents. At the age of six he was riding a horse and went straight into the branch of a tree as the horse went safely under it — the result was that he lost the sight in his right eye. Ten years later, when he was sixteen, he was fighting in a boxing match; he took a bad blow to the head and lost the sight in his left eye. And losing his sight meant that he also lost his independence. Whenever he wanted to go somewhere or do something he needed someone to help him or guide him. Even if he wanted to go to the hairdresser for a haircut someone had to take him there, and afterwards he might have to wait an hour or two before anyone was free to come and pick him up again. And there were not many jobs open to him as he grew older, for in those days most people thought that blind people could not do much for themselves.

But Morris had a fighting spirit and he was not prepared to settle for this, he was determined to lead a normal life and did not want always to depend on helpers. Then one day when he was twenty years old something happened which would change his life forever.

Some years before this there had been a terrible war, the First World War. There were thousands of young men who had been injured, and among them were a great number who had been made blind. It was then, in Germany, that guide dog training really began and, by the time Morris was a young man, it was well under way. An American living in Switzerland, Dorothy Eustis, wrote an article for an American magazine called the *Post* in which she described some of this new work that was being done with guide dogs. And so it came about that one cold and damp November day a corner newspaper-seller called out to Morris, "Hey, Mr Frank, there's a piece in this week's *Post* you ought to read! It's about blind folks like you."

Morris took the magazine home and his father read the article to him. The title of the article was "The Seeing Eye" and Morris listened with mounting excitement. As he heard of the amazing guide dogs he was filled with new hope, he knew that this was what he had been waiting for. He wasted no time in dictating a letter to Mrs Eustis. Of course he very much wanted a guide dog of his own, but right from the beginning he also had the idea of starting a guide dog training school in America. When Dorothy Eustis read Morris's letter she had to think about it carefully. Although she had visited the training school and written the article, she herself did not work with guide dogs, she trained dogs for all kinds of other jobs. But then she took a big decision; she decided to help him and wrote a letter back inviting him to come to Switzerland.

And so the day came when Morris boarded a ship to cross the Atlantic Ocean, but this was not going to be easy. Because he was blind he was not allowed to travel freely like an ordinary passenger. No, the only way he could go was to be shipped by American Express, a bit like a parcel. He was not actually wrapped up, but he had to be cared for all the time by an attendant and at nights he was locked in his room — all for his own safety of course. On the long journey he felt a bit lonely and downhearted, but he kept himself going by thinking how different it would all be when he had his own guide dog and would no longer need an attendant to lead him everywhere.

Eventually the ship docked in France and he had to go through another difficult journey before he finally arrived in Switzerland. When he stepped off the train onto the platform in Vevey he felt the warm sun and the cool air of spring, then came a pleasant voice which said, "Mr Frank, here we are."

It was Dorothy Eustis who had come specially to meet him; she shook his hand. With her was Jack, her head trainer, who also shook Morris by the hand and with a friendly laugh he said, "You're certainly the biggest parcel we have ever had from American Express!"

They were both so welcoming that Morris felt more cheerful than he had during the whole journey, and it seemed like no time at all until they arrived back at Mrs Eustis's estate. This was where Morris was going to live for the next five weeks and where the dog training took place. It had taken Morris a lot of courage to face the long and difficult journey on his own, but now it seemed as if his dreams were really going to come true.

Working together

Jack had already been training a dog for Morris; she was a German shepherd. Jack wanted Morris and his dog to be good friends right from the start, so before he introduced them he gave Morris a piece of meat. He told him that when he brought her in, Morris should welcome her, give her the meat, make a fuss over her and talk to her. Then he went to get her. Morris heard the sound of footsteps as they came into his room.

"Here girl" he said gently, and then he felt her gentle muzzle as she took the meat out of his hand. He stroked her head and ran his hand over her soft fur; he felt her wagging tail and her velvety ears. Although the dog already had a name, Morris wanted to call her "Buddy" and so, from then on, Buddy was what she was called.

The first thing Morris had to learn was how to put on the harness. There has to be a special harness on a guide dog, an ordinary lead would be no use at all because the blind person has to be in close contact with their dog. They must be able to feel at once if the dog is going left or right, if she is speeding up or slowing down, hesitating or sitting down. At first Morris was very clumsy at putting on the harness; although he was being careful, he somehow managed to poke Buddy in the eye, pinch her ear with a strap and to step on her paw. But Buddy was a patient dog and she put up with it all without complaining. Morris was keen to get on with the training, but Jack explained to Morris that Buddy still thought of him, Jack, as her master. He had trained her, and so it was very important that Buddy learn to think of Morris as her new master. And it was true, the only time she really brightened up and wagged her tail with happiness was when she saw Jack coming.

Morris knew that it was up to him to put in the work if Buddy was really going to be *his* dog. So he learned how to look after Buddy himself. It was his job to feed her, to groom her, to look after all her needs. And Buddy stayed by Morris all the time: she slept on the rug in his room, she sat under his table when he was eating, they went everywhere together. And Morris learned always to praise her and say, "Good girl" when she did well.

Then there was the first time that they worked together with the harness. Standing beside his dog Morris gave the command, "Forward!"

The handle almost jerked out of his hand and they flew along the path to the gate. Buddy stopped suddenly, Morris lost his balance and almost fell over. Jack called out, "She is showing you where the latch is."

So Morris felt for Buddy's back and on up to her nose — he found she was pointing to the latch like a teacher pointing to a letter on a blackboard. Jack was always close by calling out instructions: "Stand up straight, keep your shoulders back, hold your head high, look straight ahead, walk confidently, take long strides, hold the harness properly."

At the same time Morris had to learn to be sensitive to what Buddy was doing by paying close attention to the movements of the harness. It was no use having a guide dog unless you could learn to follow exactly where she was leading. There were so many things to remember at once. Through the handle he could feel every movement of Buddy's body and it was wonderful to be striding along. He had not gone so fast and so far for a long time.

They went down into town on the funicular railway with Jack always explaining to Morris what Buddy was doing. There were so many sounds and smells, so many commands from Jack and new things to learn. Morris began to realize what an intelligent dog Buddy was, guiding him around obstacles and keeping him safe from harm — even from overhanging branches which Buddy herself could have easily walked under. After a few more hours they got home and Morris sank into a chair. Only now did he realize how exhausted he was; his feet hurt, his legs ached, his arm was sore and he had a pain in his back from pulling against the harness. But all these aches and pains were of no importance — he felt wonderful.

This was only the beginning, for Morris needed training every bit as much as Buddy had done. Over the years he had got into all sorts of bad habits; he did not stand well but slouched, he walked hesitantly and with short steps, altogether he fumbled and lacked confidence. So during the next few days there was a lot of hard work to do and Jack often had to be very firm and strict with Morris.

Then came the day when Jack told Morris that he would now have to learn to go it alone. Jack would still be there, but he would give no more commands, there would be no more help from him — no more reminders. And so they set off for the gate. Buddy stopped but Morris was not paying enough attention and went straight into the gatepost. It really hurt, but when Jack just laughed, Morris pretended he had only brushed against the post and with a laugh he went on.

As usual, Buddy paused at the steps of the funicular railway, but Morris was too slow to respond and fell down, giving his knees a nasty bang. Jack laughed again and Morris began to feel angry inside, he thought, "What a mean way to treat a blind man." They took their seats in the cable car. Morris knew that he should check Buddy's feet were out of people's way when she lay down, but in his annoyance he forgot. Jack gently trod on Buddy's paw, just enough that she yelped. Morris jumped and hastily put Buddy under his knees, as he should have done before. Jack did not speak and Morris felt both angry and downhearted. He thought, "Why did Jack laugh at me? Why didn't he save me from falling? Why did he step on my dog to teach *me* a lesson?"

It went on like this and soon Morris was in a boiling rage. Because he was so angry, he was not paying proper attention to Buddy and he banged into people and tripped over things. When he nearly got run over by a car because of not "listening" to Buddy, it brought him sharply back to his senses, but when they got home he was still angry — angry and sad at the same time. He just said, "I don't want anything to eat." Then he went up to his room and sat on his bed and Buddy lay down on her rug.

After a while there was a knock on the door, someone opened it and Morris heard Jack's voice say, "Alright if I come in?"

Jack understood how Morris was feeling, he had known all

along that it was not going to be easy. He sat down beside Morris and spoke to him. "Look Morris, you had a hard time today, but if I go on helping you, you will never learn to depend on Buddy. You can't lean on me any more." He reminded Morris that this might be his only chance to work with a guide dog, he reminded Morris of all the blind people back home in America he could help by starting a guide dog school. The future was in his own hands.

Morris heard the door closing as Jack left the room. He felt very low. What a fool he had been to dream that everything would be wonderful. What a fool he had been to think that he could start a training school in America when he himself could not learn to work with a guide dog.

Morris felt despair and he was ready to give it all up. But then he felt someone else getting onto the bed, he felt a snuffling in his ear, he felt Buddy licking his cheek as if she too understood how he was feeling. Morris put his arm around Buddy, and then he stopped feeling so sorry for himself. Yes, tomorrow he would give it another try. Then he thought again. No, that was not quite right, he and Buddy *together* would give it another try.

Back to America

And so the training continued. Morris never again thought of giving up; from then on every day brought a bit more progress. After a few more weeks he was like a new man, with Buddy at his side he could stride out with his head held high and with a new air of confidence about him. One day he realized that he needed a haircut and so he asked someone if they could take him into town, but Mrs Eustis overheard the conversation and she said, "Morris, why don't you and Buddy go into town by yourselves? You don't need anyone to take you."

And so they went off together. They travelled into town on the railway, they went through the streets recognizing familiar places along the way, they avoided potholes and bumps, got on and off the pavements, crossed the roads and soon found the hairdressers. Morris sat down to have his hair cut and Buddy waited at his feet. Then they were ready to go home again, but there was no longer any need to rely on help from anyone else. It was another step towards independence.

By now Buddy looked on Morris as her master and Jack was careful not to interfere any more than he needed to. One day the three of them were going along the narrow road back from town and Jack stayed well behind, letting Buddy and Morris go ahead by themselves. Morris felt very happy, but then he heard a strange noise, a heavy rumble and a crashing of hooves, the noise got louder and louder until he could actually feel the ground shaking. He guessed that it was runaway horses but he had no idea how to escape. Then there came a sharp tug on the handle of the harness; Buddy dragged him violently off the road then she forcibly hauled him up a steep muddy slope at the side. The next moment he heard a team of mad, snorting animals crashing past and the sound of a heavy hurtling wagon close by. Moments later Jack came running up to them, out of breath, "Are you alright? You could have been killed." By her quick action Buddy had saved Morris from being run down.

And so the time came for Morris and Buddy to go back home to the United States. Once again Morris boarded a ship to cross the Atlantic, but the journey home was very different to the journey out. He made lots of friends because of Buddy; people would stop him and say, "What a lovely dog you have," and the conversation went on from there. Now he was free to go where he wanted on the ship, together with Buddy he could visit people until late in the evenings and then easily find the way back to the cabin.

When the ship docked in New York there was a little crowd of newspaper reporters looking for stories. Morris soon got talking and told them about all the wonderful things that Buddy could do. But the reporters were not at all convinced; one of them said to Morris, "So are you telling me that you and your dog can go *anywhere?*"

"Yes" said Morris, "Buddy can lead me anywhere."

"In that case let's see you crossing West Street," said the reporter. There was a bit of a gasp from somewhere but Morris was not put off.

"Let's go," he said.

Some of the other reporters tried to put him off because they knew that crossing West Street was far too dangerous. It was a very wide street with heavy traffic, no crossing places, and it was a famous spot for bad road accidents. Morris knew how dangerous it was, but he also knew how important it was to have the press on his side if he wanted to start a school — if he failed in this first test there might not be another chance for a long time.

Morris, Buddy and the reporters stopped at the kerb on West Street. Morris took a deep breath and gave the command, "Buddy, forward!"

Morris could feel that Buddy was looking carefully at the traffic and then she set off. The traffic was roaring, blaring and screeching, Morris could smell the hot exhaust fumes and feel the big trucks rushing close by. Angry drivers were yelling at him, but Buddy led him on, now a bit faster, now a bit slower, pausing for a moment, then going on again. The din and chaos was so bad that Morris became confused, he lost all sense of direction and was relying completely on Buddy, but Buddy carried on as calmly as if she was still in the country-side of Switzerland. After what seemed a very long time — but was really three minutes — they arrived safely on the other side of West Street. Morris knelt down and hugged Buddy close to him, "Good girl, good girl," he said, and he really meant it.

In a little while the reporter ran up to them, "You did it! I really didn't think you could. I had to take a taxi cab to get here!"

Most of the other reporters were still stuck on the other side of the street, but next day the story appeared in all the papers and then everyone in New York was keen to see this amazing dog for themselves.

After a while, Morris went home and was soon leading a full life. He had more fun with his friends than he had ever had before and he became a successful businessman selling insurance. Most clients were so pleased to welcome Buddy that they happily bought some insurance as well. Before much longer Morris was able to start his guide dog school, called the Seeing Eye.

Mrs Eustis and Jack came from Switzerland to help and soon the new school had its first two students. The Seeing Eye school is still there today and it has helped thousands and thousands of blind people, but all that still lay far in the future. At first it was difficult to find enough dogs for training. It needs a dog of a special temperament to be a guide dog. It must not be too shy nor too aggressive, it must be intelligent, calm and friendly. So only a few dogs passed the test and could be taken on for training. Finding the right people to be trainers, both of dogs and of students, was also a challenge.

Morris and Buddy travelled around the country giving lectures about the new school and big crowds of people came to see them, but, to tell the truth, many of them really came to see Buddy. In these lectures Morris and Buddy would demonstrate how they could work together. People had never seen anything like it before, and they could hardly believe their eyes as Buddy led Morris around the hall and followed his commands. It often happened that people could not believe he really was blind. Sometimes, unknown to Morris, people would put up obstacles around the lecture hall to see if they could confuse Buddy. But Buddy was never caught out, if she could not get where she wanted to go by one route, then she would always find another way round. Although it meant lots of travelling and hard work, Morris and Buddy just loved it; Morris just had to hold out her harness and Buddy would bound to him and wriggle into it all by herself.

One day Morris had some business to do in Chicago, a city he did not know, he needed to find the Gas Company building. Walking with Buddy he had been counting the blocks and thought they must nearly be there, but he was not sure and so he asked a passer-by. The man replied, "What's the matter with you? Are you blind? You're standing right in front of it!"

Before Morris could pick up the harness and give the command to go in, another man stopped him and asked, "Excuse me, can you tell me where the Gas Building is?"

Morris could not resist it, he replied with satisfaction, "What's the trouble? Are you blind? It's right in front of you!"

But they still had big battles to fight, because dogs were not allowed on planes or trains, they were not allowed in shops or restaurants, everywhere you could see signs saying, "No dogs allowed." But Morris was a fighter and never stopped working for the rights of the blind. Even though it took years of argument and persuasion, blind people with their guide dogs were eventually accepted in all these public places.

Morris and Buddy were together for ten years and there is just not enough time to tell you all the adventures they had together, perhaps one day you will read about it for yourselves. But dogs do not live as long as people do and, after a busy working life, the time came when Buddy was getting old and was no longer able to do as much. One day they were coming down the steps from a plane when Buddy's legs gave way beneath her, Morris did not want people to see that she was near the end of the road and he supported her with the harness so that nobody noticed. But when the time came to get back on board, Buddy could not make it and had to be carried up the steps. As always, there were photographers waiting to get some pictures of Morris and Buddy for their newspapers, but when they saw Buddy in her distressed state, they just lowered their cameras and gently came forward to help carry her. They were so upset to see Buddy like this that not one of them — and there were some tough characters among them — would take a picture of Buddy in her time of trouble.

Buddy did not live much longer after that. There soon came the day when she flopped down as if to say, "I have brought him home safely, I am through." She gave Morris's tear-stained face one last loving lick and fell into a well-earned sleep.

During the rest of his life, Morris had other dogs to guide him — but there was always a special place in his heart for Buddy. And so years later, when people persuaded him to write a book telling the story of his life and how the Seeing Eye began, he did not call the book *The Life of Morris Frank,* or anything like that. No, he gave his book the title, *First Lady of the Seeing Eye,* and by the first lady he of course meant his faithful helper and companion, Buddy.

Songs

Hymn

Birds fly in the air with wings out - spread. Beasts
Sweet song com- eth forth from small birds' beak. The
Strong wings car - ry birds a - bove the land. On

tram - ple the earth with hea - vy tread. But
large beasts can roar and small ones squeak. But
paws and on hoofs beasts walk, run, stand. But

up - right walks Man hol - ding high his head. Praise
Man utt - ers words, on - ly he can speak. Praise
none can com - pare with the hu - man hand. Praise

God who made all things well.
God who made all things well.
God who made all things well.

Seal Song

Waves, that roll a-long, Sing your might - y song,
Sing the song so wild and free.
Wat - ers where we roam, glid - ing through the foam,
Give us life and home in the sea.

Elephant Song

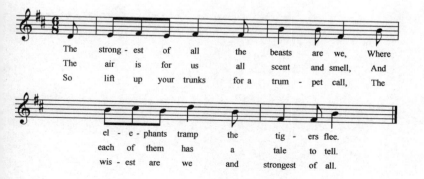

The strong - est of all the beasts are we, Where
The air is for us all scent and smell, And
So lift up your trunks for a trum - pet call, The

el - e - phants tramp the tig - ers flee.
each of them has a tale to tell.
wis - est are we and strongest of all.

Bibliography

Morris Frank and Blake Clark. *First Lady of the Seeing Eye.* Holt 1957.

Rowena Farr, *Seal Morning,* Hutchinson 1957.

Craig Holdrege, "Seeing the Animal Whole," in *Goethe's Way of Science,* State University of New York Press 1998.

Karl König, *Animals: An Imaginative Zoology,* Floris Books 2013.

Wolfgang Schad, *Understanding Mammals: Threefoldness and Diversity,* Adonis Press, 2019.

Index

Books by Charles Kovacs

Teacher resources

Class 4 (age 9–10)
Norse Mythology

Classes 4 and 5 (age 9–11)
The Human Being and the Animal World

Classes 5 and 6 (age 10–12)
Ancient Greece
Botany

Class 6 (age 11–12)
Ancient Rome

Classes 6 and 7 (age 11–13)
Geology and Astronomy

Class 7 (age 12–13)
The Age of Discovery

Classes 7 and 8 (age 12–14)
Muscles and Bones

Class 8 (age 13–14)
The Age of Revolution

Class 11 (age 16–17)
Parsifal and the Search for the Grail

General interest

The Apocalypse in Rudolf Steiner's Lecture Series
Christianity and the Ancient Mysteries
The Michael Letters of Rudolf Steiner
The Spiritual Background to Christian Festivals

Further resources for teachers

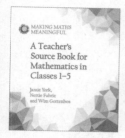

MAKING MATHS MEANINGFUL

A Teacher's Source Book for Mathematics in Classes 1–5

Jamie York, Nettie Fabrie and Wim Gottenbos

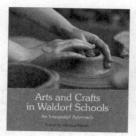

Arts and Crafts in Waldorf Schools

An Integrated Approach

Edited by Michael Martin

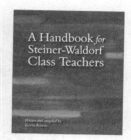

A Handbook for Steiner-Waldorf Class Teachers

Written and compiled by Kevin Avison

Painting and Drawing in Waldorf Schools

Classes 1 to 8

Thomas Wildgruber

Living Lines

Form Drawing Inspiration for Steiner-Waldorf Teachers

Henrik Thaulow

String, Straight-edge and Shadow

The Story of Geometry

Julia E. Diggins

A Journey Through Time in Verse and Rhyme

Poems collected by Heather Thomas

The Tasks and Content of the Steiner-Waldorf Curriculum

Edited by KEVIN AVISON and MARTYN RAWSON

Mathematics in Nature, Space and Time

John Blackwood

florisbooks.co.uk

Floris Books

For news on all our **latest books**,
and to receive **exclusive discounts**,
join our mailing list at:

florisbooks.co.uk

Plus subscribers get a FREE book
with every online order!

We will never pass your details to anyone else.